Studies in Computational Intelligence

Volume 680

Series editor

Janusz Kacprzyk, Polish Academy of Sciences, Warsaw, Poland
e-mail: kacprzyk@ibspan.waw.pl

About this Series

The series "Studies in Computational Intelligence" (SCI) publishes new developments and advances in the various areas of computational intelligence—quickly and with a high quality. The intent is to cover the theory, applications, and design methods of computational intelligence, as embedded in the fields of engineering, computer science, physics and life sciences, as well as the methodologies behind them. The series contains monographs, lecture notes and edited volumes in computational intelligence spanning the areas of neural networks, connectionist systems, genetic algorithms, evolutionary computation, artificial intelligence, cellular automata, self-organizing systems, soft computing, fuzzy systems, and hybrid intelligent systems. Of particular value to both the contributors and the readership are the short publication timeframe and the worldwide distribution, which enable both wide and rapid dissemination of research output.

More information about this series at http://www.springer.com/series/7092

Aleksander Byrski · Marek Kisiel-Dorohinicki

Evolutionary Multi-Agent Systems

From Inspirations to Applications

 Springer

Aleksander Byrski
Faculty of Computer Science, Electronics
and Telecommunications, Department of
Computer Science
AGH University of Science and Technology
Kraków
Poland

Marek Kisiel-Dorohinicki
Faculty of Computer Science, Electronics
and Telecommunications, Department of
Computer Science
AGH University of Science and Technology
Kraków
Poland

ISSN 1860-949X   ISSN 1860-9503 (electronic)
Studies in Computational Intelligence
ISBN 978-3-319-84637-8   ISBN 978-3-319-51388-1 (eBook)
DOI 10.1007/978-3-319-51388-1

Printed on acid-free paper

This Springer imprint is published by Springer Nature
The registered company is Springer International Publishing AG
The registered company address is: Gewerbestrasse 11, 6330 Cham, Switzerland

*To our beloved ones as a proof,
admired ones as a tribute
and adverse ones as a warning.*

Preface

For the last 40 years, a growing interest has been observed in the systems where a task to solve is decomposed into smaller parts (subtasks), which are dealt with separately and later they are synthesized into an overall solution. Such an approach may be described as *distributed problem solving*, and is usually easily implemented in parallel environments such as multi-core machines, clusters or grids. It should be noted that multi-agent systems belonging to a popular class of methods in artificial intelligence are an effective implementation of distributed problem solving. Agents are perceived as autonomous beings, which are able to interact with their environment and other agents and bear the features of intelligence. In these systems, a task to solve is usually decomposed into subtasks, which are entrusted to agents. Each agent's goal is to solve its part, and different features of agency affect this process, e.g. autonomy allows for self-adaptation of the agent's strategy.

In 1996, Krzysztof Cetnarowicz proposed an evolutionary multi-agent system (EMAS) dedicated to solving computing problems, with interesting features like distributed selection and lack of global control. Since then the idea of EMAS has been applied to different problems (e.g. single, multi-modal and multi-criteria optimization). This approach still retains high potential possibilities of extension and hybridization (e.g. with cultural or memetic mechanisms) that are researched in Intelligent Information Systems Group at AGH University of Science and Technology in Cracow, Poland.

It is noteworthy that since the inception, EMAS-related research has yielded different modification of this system (utilizing elitist, co-evolutionary or immunological inspirations). Based on these modifications, effective solutions to many difficult problems have been provided such as evolution of neural network architecture, multi-modal optimization and financial optimization to name but a few. EMAS has thus proved to be a versatile optimization mechanism in practical situations.

Multi-agent systems provide a good basis for the development of hybrid search and optimization systems, however it should be noted that in this way, more and more complex computing systems are created. Also, using common sense and remembering *Ockham's razor* rule, one should apply complex search techniques solely to difficult problems. Therefore, metaheuristics, in particular agent-oriented ones, should be treated as the *methods of last resort*, and should not be applied to simple problems.

On the other hand, the need to build complex (hybrid) systems, calls for performing a more in-depth analysis of features of their work. A detailed description of its structure and behavior is required for a full understanding of them, moreover, providing means for stochastic analysis may yield additional, important results such as confirmation if the system works at all (meaning, whether or even when it is able to localize the result).

Several formal models aimed at proving different features of evolutionary metaheuristics have been constructed. One of the first and important models of metaheuristic methods was Michael Vose's model, which proves that for a fixed size population, simple genetic algorithm (SGA) can be modeled by a Markov chain, and after further assumption that the mutation rate is positive, this chain is ergodic. This result formally justifies SGA as a well-defined global optimization algorithm. Other approaches to model evolutionary algorithms to be mentioned are different models for single-population evolutionary algorithms, proposed by Davis, Mahfoud or Rudolph. In particular, Rudolph's model was used to prove the first hitting time for a (1+1) evolution strategy optimizing a convex function. Unfortunately, there is lack of general models, as all those mentioned above are oriented on analysis of particular methods.

In order to successfully build and examine a multi-agent system, making it not only reliable, but also distributed and scalable, flexible and extensible, one must work not only on the implementation itself, but rather try to develop an universal framework, that will be later adapted by the developers creating particular flavor of the system, or even the end user himself (using eg. domain specific languages). Efficient working of such systems applied to solving of complex problem requires both high performance and flexibility connected with possibilities of adapting to dynamically changing needs of the user. Application of intelligent computational techniques supported by novel technologies makes possible adaptation of the structure and organization of the system, creating a new quality in the aspect of services provided and general applicability.

Though the main development of the agent-oriented technology occurred in connection with necessity of making possible exchange of information between cooperating and distributed agents, working in a heterogeneous environment, a significant attention is also drawn to agent-based modeling, connected with simulation-oriented applications, in particular for the process of distributed nature. At the same time, there is a lot to do in the area, which may be generally described as agent-based computing, belonging to computational-intelligence paradigm. In such systems, not the knowledge and planning of a single agent, or more complex communication protocols and interaction schemes are of utmost importance. Rather a holistic sum of single, relatively simple behaviors of individual agents affects the applicability of the whole system. From the implementation point-of-view, the situation is similar to certain simulation models, but in this case the performance becomes an important issue, as single experiments cannot always be repeated because of lack of time or hardware-related constraints.

One should be aware, that since the complexity of both computational-intelligence techniques and agent-based systems is very high, the combination of these two paradigms will pose even higher problems for modeling, design, development, monitoring the experiments and analysis of the results. In the end, design and execution of such systems demands an immense commitment and a large precision during all the stages of their realization. The systems made in *ad hoc* manner quickly become insufficient because of problems in assessment of the obtained results, that may be totally different for small changes introduced in the configuration used.

At the same time due to high complexity of the considered techniques, their applicability becomes justified not earlier as they are applied to task of relatively high complexity. The stochastic computation models used put together high exploration and exploitation capabilities, though the requirements related to the feasibility of the obtained result are satisfied after performing large number of the iterations and repetitions. Moreover, a long user-driven process of trial-and-error is often required, for proper tuning of a large number of their parameters. Performance-related issues very often make the use of classic agent-oriented plat-form useless in the case of computational intelligence applications.

Over 15 years of experience of the authors in realization of such kind of systems, also because of cooperation and involvement of many persons from different academic institutions have already resulted in preparing of many scientific publications, several doctoral theses (e.g. [91, 39, 262]) and two habilitation dissertations [40, 175], which became a basis for this monograph.

The structure of this monograph is as follows. There are three parts, first one is devoted to the literature review, motivation and definition of the considered systems, also including full formal analysis of EMAS leading to the conclusion that the computing based on this paradigm is formally justified, as such systems are always able to locate the solution to be found. The second part is devoted to design and implementation of the platforms supporting EMAS-like computations, along with the presentation of the AgE platform, that was used during the computations which results are presented in this book. The last part is fully devoted to the experimental results obtained by applying EMAS and some of its modifications to solve discrete and continuous problems, exploring the possibilities of adaptation of particular parameters of the system to solve benchmark problems of varying difficulty.

Chapter 1 presents a systematic state-of-the-art review. It begins with discussion on features of computing systems and their relation to decision support, including identification of difficult problems (so-called "black-box") search problems and justification for the use of complex metaheuristics to solve them. Later, evolutionary and hybrid metaheuristics are reviewed and certain gaps are identified in order to prepare the reader for the next chapters.

Chapter 2 gives a description of agent-based architectures of computing systems is given. Then, a concept of Evolutionary Multi-agent Systems (EMAS) is discussed, including base mechanisms that are used for control and tuning up of its work, along with physical and logical structure of the system that gives the starting point for later architecture-related deliberations. Besides classic EMAS, several of its variations are also presented leveraging different natural inspirations (immunological, co-evolutionary or memetic).

Chapter 3 starts with the formal definition of EMAS and construction of a Markov chain modeling dynamics of EMAS. Then, after necessary assumptions, the ergodic theorem is defined and a full formal proof is given. Later, the same structure of presentation is retained for iEMAS, however, in this case no full formal proof has been constructed, and only necessary conjecture is formulated and the proof outline is described. The chapter is concluded with a short description of actual goals reached in the formal analysis of agent-based metaheuristics.

Chapter 4 builds a technological perspective for the considered class of systems by introducing the most important concepts, ideas and techniques in the field of agent-based systems and component technologies, pointing-out possible relations between these approaches on the development level.

Chapter 5 is opened with the discussion of potential requirements connected with managing of the system, considering its dynamic nature, both because of potential changes of the task to be solved (non-stationary problems), user-related requirements or preferences and the structure of the computing environment. An evaluation of the existing tools was also conducted, in order to check their potential capabilities in realization of agent-based computing systems. Referring to the proposed architectural model and showing the design assumptions *de facto* a new method of implementation of agent-based systems for computational applications was presented.

The most important components of the referential implementation of the agent-based platform and distributed computing environment AgE are described in Chap. 6. Particular attention was given to the previously announced technological requirements and relations to alternative tools described in the literature.

Chapter 7 presents the results of a wide-ranging series of experiments conducted in order to evaluate the efficiency of agent-based metaheuristics, as compared to classical search methods. In the beginning, EMAS is evaluated using selected high-dimensional benchmark functions. After presenting the benchmarks, comparison between EMAS and PEA (parallel evolutionary algorithm) is made, using classical (evolutionary) and memetic versions of these methods. Later, immunological version of EMAS is tested versus the classical EMAS.

The final Chap. 8 focuses on, tuning of selected EMAS parameters is considered. After observing an impact of changing certain EMAS parameters (energy-related and probabilistic), and iEMAS (lymphocyte parameters), the results are summed up, which provides a base for further use in order to adapt these metaheuristics to particular problems.

Kraków, Poland   Aleksander Byrski
August 2016   Marek Kisiel-Dorohinicki

Contents

Part I
Concept and Formal Model

Chapter 1
Contemporary Methods of Computational Intelligence

When facing lack of algorithms suited for many important class of computing problems, promising results can be obtained using heuristic-based techniques, utilizing different nature-oriented, physical or social inspirations. Such algorithms are very often characterized as belonging to of *computational intelligence* methods [104, 251], because of its broad applications possibilities and potential adaptation capabilities.

Their characteristics very often are strongly dependent on the solved problem and on particular run (they often rely on stochastic techniques), they are also very often called *soft computing* methods [31].

A significant class of these methods is based on natural or social inspirations (called *population techniques*), and are very often applied to solving optimization problems. One of their characteristic features is a huge potential of parallel processing, that is especially important for their efficiency, because of their intrinsic stochastic nature [280]. The most popular methods belonging to this class are the ones inspired by the rules of natural evolution, i.e., **evolutionary algorithms** or *evolutionary computation*. Other popular optimization methods are *artificial immune systems*, (ang. *memetic algorithms*) [219], *particle swarm optimization* [164], *ant colony optimization* [87] and many others.

The current chapter presents a short introduction into the topic of *population-based techniques of computational intelligence*, and survey of selected concepts important in this field, especially applied to search and optimization. The most important topics tackled in this chapter are evolutionary computing and theirs parallel models, considering their popularity and importance for the deliberations presented in the subsequent chapters.

© Springer International Publishing AG 2017
A. Byrski and M. Kisiel-Dorohinicki, *Evolutionary Multi-Agent Systems*,
Studies in Computational Intelligence 680, DOI 10.1007/978-3-319-51388-1_1

1.1 Computing Systems and Decision Support

Virtually all areas of human activities are connected with decision undertaking or choosing different proceeding strategy, that will realize the assumed goal (or a set of goals) in the best possible way. Many decision problems can be described by a parametric mathematical model, for which a dedicated quality function may be formulated. Then, the decision problem is equivalent to search for such values of its parameters (decision variables) that minimize (or maximize) a certain goal function (quality factor). In other words it consists in search in space of potential solutions, in order to find an *optimal* variant. In particular, many technical or economical issues can be formulated as *optimization* problems.

1.1.1 Formulation of Optimization Problem

Particular considered optimization problem describes in the first place the *search space*, that in a general case may be any metric space $(U, | \cdot |)$. In the *parametric optimization* tasks, each point $x \in U$ is a vector of independent variables (so called *decision variables*), that can take discrete $(U = \mathbb{Z}^n)$ or continuous $(U = \mathbb{R}^n)$ values, while n (the number of decision variables) is equal to so called *dimensionality* of the task, which practically is the most important factor showing its difficulty level.

The set of feasible solutions $D \subseteq U$ can be equal to the whole search space, however usually it is its subset. In such case the problem is described as optimization *with constraints*, that can be formulated as the following conditions:

- inequality constraints: $g_k(x) \geq 0, k = 1, 2, \ldots, K$,
- equality constraints (so called bonds): $h_l(x) = 0, l = 1, 2, \ldots, L$.

Assuming, that the quality factor (called the goal function) is described by the following expression: $f : U \to \mathbb{R}$, the minimization task can be formulated as search for such $x^* \in D$ that (see e.g. [150]):

$$x^* = \arg \min_{x \in D} f(x). \tag{1.1}$$

Of course the maximization task can be easily adapted to the minimization task by simply changing the sign of the goal function.

The optimization task becomes much simpler when it is known, that the goal function has only one extremum in the feasible search space. For the continuous tasks, such assumption guarantees that the goal function is convex. Otherwise, the goal function is *multi-modal*, and can have many local extrema (i.e., points, which have certain, non-empty neighborhood, not containing better solutions) in the feasible search space. For the minimization task, it can be formulated as follows:

$$\exists \delta > 0 \; \forall r < \delta \; \forall y \in N_r(x) \cap D \; f(x) \leq f(y) \tag{1.2}$$

where $N_r(x)$ is a neighborhood of x with a radius r, according to the metric used in the space U.

In such tasks, one can realize so-called *multi-modal* optimization, focused on finding as many as possible of the local extrema, or *global* optimization, where only one, globally optimal solution is sought [8].

The optimization problem becomes quite complex, when the quality is expressed by many irreducible factors (poly-optimization, multi-criteria optimization). In such case even the formulation of the task becomes non-trivial, as it requires a definition of certain order relation in (multi-dimensional) criteria space. It seems that the most natural assumption would be to take (weak) ordering of vectors—*Pareto approach, Pareto optimality* [267, 273]. The relation defining such order is usually called a *dominance relation*. In the task of minimization of a function vector $F = [f_1, f_2, \ldots, f_M]^T$ (so called criteria vector), where $f_i : \mathbb{R}^N \rightarrow \mathbb{R}$ it can be said that the solution x^a *dominates* the solution x^b if and only if:

$$\forall m = 1, 2, \ldots, M \ \ f_m(x^a) \leq f_m(x^b) \ \text{ and } \ \exists m \ \ f_m(x^a) < f_m(x^b) \tag{1.3}$$

The multi-criteria optimization (in the Pareto sense) consists in finding non-dominated solutions (in other words *effective* solutions or *Pareto-optimal* solutions) in the set D. Such concept of optimality, by design does not lead to finding a single solution, but rather a set of solutions (so called Pareto *set* or *front*), which later can be analyzed in order to choose a certain solution [63, 78, 228].

1.1.2 Decision Support and Computational Intelligence

The *optimization* itself is a process leading to choosing one, the best from the point of view of the quality factor (criterion) solution (decision), from a certain set of possible (feasible) variants. Of course, in practical decision support cases, it is very often acceptable to find a *good enough* solution, or simply a solution better than the one that is already known (of course during an acceptable period of time, from the user's point of view).

In such cases, *heuristic* methods can be efficiently used. Such methods turn out to be especially well-suited when exact mathematical models of the tackled problems are not know, no deterministic numerical methods exist or their computational complexity is too high. Such techniques usually do not guarantee obtaining of exact solutions, but rather certain estimates, that can be *good enough* for many practical applications. Among such techniques, a prominent position is occupied by the natural-inspired techniques that belong to the field of *computational intelligence*.

Each search heuristics consists of three components: a solution generator, a selection method and a stopping condition. Generally speaking, heuristics may be described as not fully-fledged search algorithm, which delivers in any time moment a certain candidate solution. Such algorithm processes iteratively one or more variants of the decision (so called *working set*), creating based on these new variants (solution generator) and choosing based on their evaluations these, that will become the next working set (selection rule), as long as the stopping criterion is satisfied. Because each of the heuristics parts can contain stochastic rules, there is no guarantee that increasing the number of iterations will hep in finding a better solution [278]. Moreover, choosing of an algorithm to particular problem requires experimental evaluation of efficacy, which is usually realized by multiple running of the same algorithm with different configurations of its parameters. Choice of parameters values can be automatized by using another optimization method (meta-optimization), however in this case one must care for defining of quality factor showing the efficiency of the base method. Alternatively it is possible to adapt the parameters during the run of the algorithm, but in this case the algorithm for adaptation must be chosen appropriately and its parameters must be set.

Unfortunately, modeling of such methods is a difficult and unrewarding task, though there exist formalisms that can be used for estimation of certain aspects of some algorithmic classes, mostly applied for evolutionary techniques (see different models of single-population algorithms, e.g. [71, 134, 197, 248] and multi-population ones, e.g. [55, 109, 280]). The most interesting results were however obtained for models describing genetic based on stochastic processes, which can be used for showing the algorithm work from the perspective of asymptotic convergence in the probabilistic measure space [224, 253, 291]). However, significant results were obtained mostly for simple configurations, proving certain practical features of the computation, such as e.g. *first hitting time* (see [24, 143]). There exist also a number of extensions of Vose's model, prepared for hybrid metaheuristics (see, e.g. [50, 180, 255]). However, formal proofs of asymptotic features are rare, with a noteworthy exclusion of full ergodicity proof for parallel genetic algorithm (treated as direct extension of Vose's model) [256]. The practical applications of such proofs is quite limited, as many assumptions must be made in order to make these proofs work. In particular it is impossible to show a particular form or parametrization of the modeled method for any considered problem [14].

1.1.3 Difficult Search Problems

Michalewicz and Fogel in [212] propose several reasons why the problem may be considered *difficult*, e.g. the number of possible solutions is too large to perform an exhaustive search for the best answer; the problem is so complex that in order to provide any feasible answer, a simplified model must be used; the evaluation function describing the quality of the solution is noisy or varies with time and therefore many solutions are sought.

Certain search problems, which fall into the description given above, are perceived to be difficult *per se*, because their domains are very hard or even impossible to be described and explored, using conventional analytical methods (see, e.g. combinatorial optimization problems [229]). The setting of such problems is sometimes called "black-box scenario" [102].

According to a definition given in the previous section, let us assume that there exists a meta-algorithm covering all randomized search heuristics working on the finite search space $\mathcal{D}$. Functions to be optimized are all functions that may be described as $f : \mathcal{D} \rightarrow [0, M]$. Now the "black-box" scenario is defined as follows [102, Algorithm 1].

1. Choose some probability distribution p on $\mathcal{D}$ and produce a random search point $x_1 \in \mathcal{D}$, according to p. Compute $f(x_1)$.
2. In step t, stop if the stopping criterion is fulfilled. Otherwise, depending on the up-to-date candidate solutions $I(t) = (x_1, f(x_1), \ldots, x_{t-1}, f(x_{t-1}))$, choose some probability distribution $p_{I(t)}$ on $\mathcal{D}$ and produce a random search point $x_t \in \mathcal{D}$ according to $p_{I(t)}$. Compute $f(x_t)$.

If a certain problem can be solved with this scenario only (in a reasonable time), it can be called a "black-box problem". In other words, this notion encompasses all the problems, whose candidate solutions may be sampled randomly, but there is no means of deriving them on the basis of the existing knowledge of the search space $\mathcal{D}$. To sum up, such problems may only be solved using general-purpose algorithms (i.e., heuristics), taking into consideration little, or no information from a problem domain

Randomized algorithms (those using random, or pseudo-random choices) are usually classified as Monte Carlo (because they provide an approximate solution) or Las Vegas (because they finally provide a correct solution, if enough time is given) algorithms [10].

Unfortunately, there is no guarantee that heuristics will find satisfactory solutions, therefore, their features observed for particular problems must be verified empirically. This is simply because theoretical analyzes take a number of assumptions of the algorithm (that is inevitable when constructing a simplified model of reality). On the other hand, this simplification may hamper the applicability of the model in real-world scenarios, therefore an experimental verification is required to make sure that both heuristic and its model are valid.

Yet a heuristic may often give an approximate solution with controllable adequacy, which means that a process solving can be stopped by a decision maker once he is satisfied.

Although these algorithms process certain points from the problem domain, having at the same time complete information about the value of the criteria function of this particular solution, the global features of the search, such as e.g. the information about closeness to the optimum, remain hidden. The whole search process consists in more or less complex iterative sampling of the problem domain.

Complex approaches that may be used to solve such difficult problems (see, e.g. [4]) somehow relieve the user of a deep understanding of intrinsic relations among the different features of the problem itself, instead constituting "clever" and "general" computing systems. No one can claim that the Holy Grail of search techniques has been found, thinking about these universal techniques, as well-known "no free lunch theorem" must be kept in mind. Wolpert and Macready prove that all search and optimization techniques are statistically identical when compared for all problems described in certain domain [296, 297]. So there is still much to be done to adapt parameters of these techniques to solve certain problems.

1.1.4 Metaheuristic and Heuristic Search Methods

A general definition of a heuristic algorithm, without giving details such as particular problem, accurate definition of search space or operators is called a *metaheuristic*. In this way, a simple heuristic algorithm, such as greedy search may be defined as, e.g. "iterative, local improving of a solution based on random sampling", without going into details of the nature of random sampling or the explored space. Therefore, metaheuristics are usually defined as general purpose, nature-inspired search algorithms [130].

Blum and Roli [30] provide a summary of the metaheuristic properties:

- they are approximate and usually non-deterministic,
- their goal is to efficiently explore the search space seeking for (sub-)optimal solutions,
- they "guide" the search process,
- they may incorporate mechanisms dedicated to avoiding being trapped in local extrema,
- they are not problem-specific,
- they can utilise search experience (usually implemented as some kind of memory mechanism) to guide the search.

Such techniques are usually nature-inspired and follow different phenomena observed in e.g. biology, sociology, culture or physics.

Another type of heuristic algorithms are so-called hyper-heuristics, which utilize more advanced mechanisms (e.g. from the domain of machine learning) to optimize the parameters of the search, or even select an appropriate lower-level search method [37].

A simple, but effective classification of metaheuristics (cf. [36, 88]), which gives a sufficient insight into the problem for the purpose of further considerations, is as follows:

- Single-solution metaheuristics work on a single solution to a problem, seeking to improve it in some way. The examples are local search methods (such as local-search, greedy heuristic, tabu search or simulated annealing) [278].
- Population-based metaheuristics explicitly work with a population of solutions and put them together in order to generate new solutions. The examples are evolutionary algorithms [133], immunological algorithms [69], particle swarm optimization [163], ant colony optimization [86], memetic algorithms [218] and other similar techniques.

Most of this book is focused on population-based metaheuristics, however, below selected popular single-solution metaheuristics will be presented in short, in particular with regard to the creation of hybrid systems.

Stochastic Hill Climbing

One of the simplest and most popular single-solution metaheuristics is the stochastic hill-climbing algorithm (see Pseudocode 1.1.1). Starting with a random solution, the algorithm selects subsequent ones (also randomly) and accepts them only if they improve the value of a certain predefined goal function. The algorithm was primarily designed to solve combinatorial optimization problems (for classical implementations refer to, e.g. [117, 159, 215]), however, it can be easily used in continuous optimization by appropriately defining the search step. This algorithm may be easily incorporated into hybrid systems (e.g. fulfilling the role of local-search algorithm in memetic computing) [292].

Pseudocode 1.1.1: PSEUDOCODE OF HILL CLIMBING ALGORITHM

$current \leftarrow RandomSolution()$
for $i \in [1, max]$
　　　　⎡ $candidate \leftarrow randomNeighbour(current)$
do ⎨ **if** $goalFunction(candidate) \geq goalFunction(current)$
　　　　⎣ **then** $current \leftarrow candidate$

Simulated Annealing

A more sophisticated approach to random local search is simulated annealing. Inspiration for this algorithm comes from an annealing process in metallurgy. In this process, a material is heated and slowly cooled under specific conditions in order to increase the size of the crystals in the material and to reduce possible defects that may arise in the cast. Using this metaphor, each solution in the search space is treated as a different value of internal system energy.

　　The system may be heated (in this case the acceptance criteria of the new samples are relaxed) or cooled (in this case the acceptance criteria are narrowed). Once the system is cooled down, the final suboptimal solution is obtained. To sum up, in this algorithm, the search space is probabilistically re-sampled, based on Metropolis-Hastings algorithm [142] for simulating samples from a thermodynamic system [168] (see Pseudocode 1.1.2). This algorithm was also designed for combinatorial optimization, but may be easily adapted to continuous problems [192].

Tabu Search

Tabu search belongs to the class of global optimization metaheuristics, however, it may be easily used for controlling an embedded heuristic technique (creating a hybrid search method). It is a predecessor of a large family of derivative approaches which introduce memory structures into metaheuristics.

Pseudocode 1.1.2: PSEUDOCODE OF SIMULATED ANNEALING ALGORITHM

$current \leftarrow randomSolution()$
$best \leftarrow current$
for $i \in [1, max]$
do $\begin{cases} S_i \leftarrow randomNeighbour(current) \\ temp \leftarrow rcalculateTemperature(i, tempMax) \\ \textbf{if } goalFunction(S_i) \leq goalFunction(current) \\ \quad \textbf{then } \begin{cases} current \leftarrow S_i \\ \textbf{if } goalFunction(S_i) \leq goalFunction(best) \\ \quad \textbf{then } best \leftarrow S_i \end{cases} \\ \quad \textbf{else if } (exp(\frac{goalFunction(current)-goalFunction(S_i)}{temp}) > random()) \\ \quad \textbf{then } current \leftarrow S_i \end{cases}$

The main goal of the algorithm is to help the search process avoid returning to recently visited areas of the search space (cycling). The method is based on maintaining a short-time memory of the recent solutions visited in the course of search, refusing to accept the new solutions which are the same (or close) as the ones contained in the memory (see Pseudocode 1.1.3).

The algorithm was introduced by Glover and applied to optimization of employees duty roster [131] and Traveling Salesman Problem [129]. Nowadays it is one of the most popular algorithms hybridized with other search techniques (see, e.g. [160, 204]).

Pseudocode 1.1.3: PSEUDOCODE OF TABU SEARCH ALGORITHM

$best \leftarrow RandomSolution()$
$tabuList \leftarrow \emptyset$
while **not** $stoppingCondition()$
do $\begin{cases} candidateList \leftarrow \emptyset \\ \textbf{for } (candidate \in neighborhood(best)) \\ \textbf{if } (\textbf{ not } contains(tabuList, best)) \\ \quad \textbf{then } candidateList \leftarrow candidateList \cup \{candidate\} \\ candidate \leftarrow locateBestCandidate(candidateList) \\ \textbf{if } (goalFunction(candidate) \leq goalFunction(best)) \\ \quad \textbf{then } tabuList \leftarrow tabuList \cup \{candidate\} \\ best \leftarrow candidate \\ trimTabuList() \end{cases}$

1.2 Evolutionary Metaheuristic Techniques

Even though the beginning of research on evolutionary methods can be dated in the late 50s of XX century (e.g. works of Box [33] or Friedberg [120, 121]), this field was quite unknown to a broad scientific community for a long time. It seems that hardware-related constraints [112] and flaws of the first methods [14] were the main causes for this.

The situation began to change in 60s, when several seminal works for this field were completed by Holland [148], Fogel [113], Rechenberg [239] and Schwefel [260]. These works were not acclaimed in the beginning and they were treated with a huge dose of skepticism, however as it turned out later, they became starting points for the following strongly interconnected, though for a long time developed independently, approaches to the evolutionary computing:

- John Holland [147] in 1975 modeled the process of evolution of the individuals constructed with the use of binary code. He was the first researcher to utilize predefined operators used to change genotypes, which were similar to crossover and mutation. He found out that the average fitness of this population tends to increase. A similar algorithm under the name of **genetic algorithm** was later popularized by David Goldberg [133].
- Ingo Rechenberg [240] and Hans-Paul Schwefel [259] researched optimization of mechanical devices by permuting randomly-generated solutions. Having observed certain similarities to the biological evolution process in their approach, they invented methods known under the name of **evolution strategies** [258].
- Lawrence Fogel [114] tried to model the process of inception of artificial intelligence upon an approach based on self-organization. He evolved finite automata aimed at understanding a predefined language [113]. This approach was called **evolutionary programming**, and after further adaptation became a popular technique in optimization [114].
- John Koza tried to work on automatic generation of computer programs using evolutionary algorithms. His research focused on evolving LISP program structures using a tree-based encoding, which is natural for this language. In this way, a technique called **genetic programming** was devised [181].

Since the beginning of 80s the number of publications and conferences tackling particular evolutionary techniques increases cf. [2]. However, during 90s, a real stratification of concepts connected with different evolutionary techniques or evolutionary algorithms can be observed (probably because of start of publishing of Evolutionary Computation journal in 1993). A detailed survey of evolutionary techniques may be found in [15].

1.2.1 Biological Inspirations

The origins of the evolutionary algorithms may be found in the 19th-century works of Gregor Mendel—the first to state the baselines of heredity from parents to offspring—who demonstrated that the inheritance of certain traits in pea plants follows particular patterns (now referred to as the laws of Mendelian inheritance). Later in 1859, Charles Darwin formulated the theory of evolution [68].

Evolutionary algorithms are based on analogies to natural evolution, which can be observed as changes in diversity of species in generations [12, 112]. The evolution is connected with the ability of organisms to reproduce and die, however these phenomena do not clarify the arising and disappearing of certain characteristics in subsequent generations. According to Darwinian Evolution Theory, this phenomenon can be explained based on *natural selection* and *mutation*.

Natural selection allows to survive the species that are the best fit to the current environment conditions. This process is related to constraining of the population size by the limited resources. As a result, the highest chance for survival belongs to the organisms that use the available resources in the optimal way. Moreover, in subsequent populations, small, random changes of the individuals can be spotted (mutations). If such changes turn out to be beneficial in the current environment, they will be amplified by the selection mechanism. Otherwise, they will disappear in the next populations.

The description above is an acclaimed point of view for explaining evolutionary processes in the macro-scale [72]. Contemporary *synthetic evolution theory* (sometimes called a neo-darwinism) extends this model with molecular-level inheritance mechanisms. This theory is based on the concept of *gene* as a basic inheritance unit. During the reproduction, a complete set of genes (*genotype*) is passed to the offspring, subjected to recombination and mutation first. Based on the inherited genetic information and the features of the environment, the individual characteristics (*phenotype*) is developed. The natural selection tackles the individuals in the population, thus it is connected with phenotypes. Random changes in the genetic information occurring during the inheritance as an effect of mutation and recombination will be evaluated based on the phenotype (affected by the environment too).

Contemporary perception of the evolution processes, assumes as its basic units *genes*, as selection units *individuals*, and as evolution unit: *population* consisting of individuals, or in other words, a pool of genes contained in them.

1.2.2   Structure of Evolutionary Algorithm

Evolutionary algorithm consists in processing of *population of individuals* representing the set of potential solutions of a given problem. Evolution of population of solutions is realized as producing of subsequent generations using so-called *genetic operators* (or *variation operators*) and the *selection* process. The aim of the genetic operators is to introduce random variations into (by the means of *mutation*) and to exchange (*recombination*) the genetic material of the individuals, resulting in search for new solutions. The selection of the best individuals is realized based on the *fitness function* provided by the environment—being a way of measuring the quality of the selected individuals' solutions. Thus the evolution process should strive for generating better and better individuals and to find the (usually approximated) solution of the given problem. As an final solution, the best individual generated in any moment of the evolutionary search is considered.

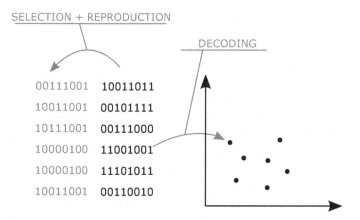

Fig. 1.1 Genotype and phenotype space in evolutionary algorithm

The search space of the evolutionary algorithm consists of a set of feasible codes I, sometimes called *genotype space* [12]. However, for the real-world problems, the search space is usually a set of objects, with many characteristic parameters of different type $x_i \in D$ (See Fig. 1.1). In such case, it is necessary to use a coding function, making possible phenotype–genotype mapping:

$$h : D \to I \tag{1.4a}$$

and otherwise:

$$h' : I \to D \tag{1.4b}$$

Based on these parameters it is possible to define the fitness function, that becomes a basis for the definition of the selection process:

$$f : D \to R \tag{1.5}$$

In the optimization problems, the fitness function is the optimized goal function, however, depending on the adopted selection model, it can require scaling in order to achieve satisfactory diversity level of the individuals in the population (so called selection pressure) [210].

I the population of μ individuals $a_i \in I$ is denoted by P, and these individuals represent all possible solutions of the problem to be solved, then in the next step t of the algorithm:

$$P[t] = \{a_1^t, \ldots, a_\mu^t\} \tag{1.6}$$

The initial population $P[0]$ is usually created randomly. In justified cases, however, it might make sense to point out a certain starting point — e.g. when the evolutionary algorithm is only a part of a complex, usually hybrid, computing method.

The variation operators used during reproduction may be *asexual*, when the new code is generated based on exactly one parent, after introducing small random changes into the genetic material (mutation) or *sexual*, when the new code is generated based on the genetic material of two parents (recombination), usually exchanging parts of their codes (crossover). The number of the individuals in the offspring population can be different, based on the adopted selection model:

- for the *steady state* evolutionary algorithm $\lambda = 1$,
- when $1 \leq \lambda \leq \mu$, it is possible to choose the value of the *generation gap*,
- in nature, it is normal to have a *reproduction surplus*, namely $\lambda > \mu$.

The selection is based on choosing μ individuals from the parent population (preselection) or offspring population $P'[t]$ and an additional set of individuals Q (postselection). However the most common situation is when $Q = P[t]$ or $Q = \emptyset$ in so called *generational* evolutionary algorithm.

One of the immanent features of evolutionary algorithms is asymptotic convergence (the probability of reaching the optimal solution converges to 1 as time tends to infinity). There is a lack of mathematical proofs of this feature for popular metaheuristics, however, Michael Vose presented a detailed proof of asymptotic convergence for a simple genetic algorithm modeled with the use of Markov chains [290].

The two main features of evolutionary algorithms are their random nature and the fact that they produce suboptimal results. Both these features lead to an important conclusion: the user is unsure whether the solution found recently is close to the optimal. He also does not know, when (and if) the optimal solution will be generated. Therefore, an appropriate definition of stopping criteria is crucial for these search techniques.

Practical stopping criteria seldom utilize detailed the analysis of the search process, instead they consider the basic features of the observed generations [8, 133]:

1. Monitoring the solutions generated by the algorithm based on the phenomenon that at the beginning of the search better individuals (compared to the current population) are created more frequently than later, e.g.:

    - Criterion of the *maximal cost* based on the assumption that if a certain cost reaches a predefined value, the algorithm is stopped (e.g. after checking that the maximal number of generations has been reached).
    - Criterion of *satisfactory fitness function level*: the algorithm is stopped, when the best individual crosses a certain, predefined value of the fitness function. The application of this criterion is somewhat risky, because the user must set arbitrary level of the fitness function when its properties are unknown. Moreover, none of the individuals may cross the predefined fitness value, so in the worst case the algorithm can run infinitely.

- Criterion of a the *minimal improvement speed*, based on the observation of the results produced by the algorithm which is stopped when there is no further improvement in the solution in the predefined period of time. This criterion is also risky as the algorithm may get stuck in a local extremum of the fitness function for a longer time; so again, the accurate choice of the mentioned period of time should require the knowledge of the properties of the fitness function in order not to stop the search prematurely.

2. Monitoring exploration features of the algorithm. These criteria are based on experimental observation of the loss of diversity, e.g.:

- Criterion of *loss of population diversity* based on computing a certain diversity measure and stopping the search when its value falls below a certain level. This approach is based on the assumption that the algorithm, having passed the exploration phase, starts exploitation when new individuals are generated near a certain extremum of the fitness function (not necessarily the global one).
- Criterion of the *deterioration of the self-adaptive mutation operator* based on the experimentally proved hypothesis that the mutation range adaptation tends to decrease in the evolutionary algorithms (this is again caused by passing into the exploitation phase).

The easiest of these criteria are of course the cost ones (based on generation count, or on the overall time passed from the beginning of computation).

1.2.3 Avoiding the Local Extrema

Solving difficult search and optimization problems (e.g. black-box ones) introduces additional requirements for the evolutionary algorithms concerning the ability to avoid or escape from the local minima. This feature is crucial to achieve a balance between the most important features of search techniques, namely *exploration* and *exploitation* [133, 210].

Exploration, as defined by March "...includes things captured by terms such as search, variation, risk taking, experimentation, play, flexibility, discovery, innovation" [199] while exploitation "...includes such things as refinement, choice, production, efficiency, selection, implementation, execution" [199]. In terms of metaheuristics, exploration is the ability to conduct broad search in every part of the admissible search space, in order to provide a reliable estimate of the global optimum, whereas exploitation consists in refining the search to produce a better solution [277].

In population-based methods, retaining diversity is an outcome of a balance between exploration and exploitation abilities (cf. classical evolutionary algorithms discussed in a textbook by Michalewicz [210]). An extensive survey of exploration and exploitation balance retaining methods for evolutionary algorithms is given in [65], however, this point of view may be easily extended to all population-based metaheuristics.

Referring to selected classical diversity enhancement techniques, several decomposition and coevolutionary techniques come in mind. *Niching* (or *speciation*) techniques [198] are aimed at introducing useful population diversity by forming subpopulations (also called "species"). *Allopatric* (or *geographic*) *speciation* may be considered when individuals of the same species become isolated due to geographical or social changes. *Decomposition* approaches of so-called *parallel evolutionary algorithms* (PEA) model such phenomena by introducing non-global selection (mating) and some spatial structure of population [54].

In a *coarse-grained* PEA (also known as *regional* or *multiple deme* model), the population is divided into several subpopulations (regions, demes) and selection is limited to individuals inhabiting one region, and a migration operator is used to move (copy) selected individuals from one region to another. In a *fine-grained* PEA (also called a *cellular* model) individuals are located in some spatial structure (e.g. lattice) and selection is performed in the local neighborhood.

In *coevolutionary* algorithms, the fitness of each individual is not computed directly based on the definition of the problem to be solved, but results from interactions with other individuals residing in the population. In *cooperative* coevolutionary algorithms, a problem to be solved is decomposed into several subproblems solved by different algorithms in separate subpopulations [236]. Cooperation between individuals from different subpopulations may take place only during a phase of computing fitness for the complete solution. The fitness value is computed only for the group of the individuals from different subpopulations, which form a complete solution to the problem.

In *competitive* coevolutionary algorithms usually two individuals compete with each other in a tournament and their "competitive fitness" corresponds to the outcome of this competition [230]. In each algorithm step, a given individual from one subpopulation competes with its opponents taken from other sub-populations. The results of this competition have an impact on the current fitness of the individual which mates with partners coming from the same subpopulation. This mechanism can be applied irrespectively of the number of subpopulations used in the algorithm—it can be used even if there is only a single population. In this case, opponents are chosen from the same population.

In evolutionary search methods, when taking into consideration that many solutions are processed at the same time with variation operators (e.g. crossover and mutation in evolutionary algorithms), maintaining population diversity is crucial. Lack of diversity leads to stagnation and the system may focus on locally optimal solutions (in other words—trapped in a local extremum), lacking the diversity to escape [209]. Therefore, providing appropriate means of measuring and retaining population diversity is a very important task, which in this monograph is measured according to the following two methods:

- Morrison–De Jong (MOI) measure based on the concept of moment of inertia for centroid (centre of gravity computed for points distributed in multi-dimensional space) [217]. This measure is closely dependent on the distribution of the individuals across the search space.

- Minimal standard deviation (MSD) of each gene computed for all individuals in the population, similar to the column-based method proposed by De Jong [76]. This simple measure focuses on dispersion of the average values computed for individual genes.

It is easy to see that MOI measure tends to be more general than MSD, therefore, observing both measures may yield interesting results that show different aspects of population diversity.

1.2.4 Parallel Evolutionary Algorithms

Evolutionary algorithms, as an abstraction of natural evolutionary processes, are apparently easy to parallelize and many models of their parallel implementations have been proposed (cf. [3, 55]). The standard approach (sometimes called a *global parallelisation*) consists in distributing selected steps of the sequential algorithm among several processing units. In this case a population is unstructured (*panmictic*) and both selection and mating are global (performed over the whole population).

Decomposition approaches are characterized by non-global selection/mating and introduce some spatial structure of a population. In a *coarse-grained* PEA (also known as *regional* or *multiple-deme* model) a population is divided into several subpopulations (regions, demes). In this model selection and mating are limited to individuals inhabiting one region and a migration operator is used to move (copy) selected individuals from one region to another. In a *fine-grained* PEA (also called a *cellular* model) a population is divided into a large number of small subpopulations with some neighborhood structure. Here selection and mating are performed in the local neighborhood (overlapping subpopulations). It is even possible to have only one individual in each subpopulation (this is sometimes called a *massively* parallel evolutionary algorithm).

And finally there are also methods which utilize some combination of the models described above, or even a combination of several instances the same model but with different parameters (*hybrid* PEAs).

Global Parallelisation

In PEA with a global population selected steps of the sequential algorithm (mostly evaluation of individuals) are implemented in a distributed or multiprocessor environment.

In a *master-slave* model one processing unit (*master*) is responsible for management of the algorithm and distribution of tasks to other processing units (*slaves*). This often consists in sending selected individuals to slaves to perform some operations (e.g. calculate fitness). In a *sequential selection* model a master processing unit waits for finishing computation in all slave nodes so there is a clear distinction between successive generations. This makes the implementation simple, yet the cost of idle time of slaves is a potential bottleneck of this method. In a *parallel selection* model

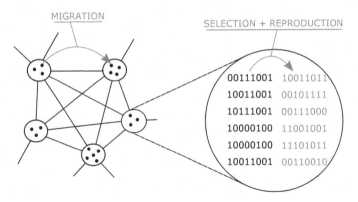

Fig. 1.2 Parallel evolutionary algorithm with migration

a master processing unit does not wait for all slaves but when one finishes the work it is immediately allocated a new task. In this case selection may be done in several variants, for example in form of a tournament (so one needs to know only a subset of fitness values).

Conversely, in a system with *shared memory* all processors have access to all individuals. This approach has all advantages of the master-slave one without its handicap, yet requires some kind of control over possibly simultaneous operations on individuals.

Regional Models

Regional models introduce coarse-grained spatial structure of a population, suitable for implementation in a distributed architecture. Each node has its own (sub)population and runs a separate thread of evolution (Fig. 1.2), thus conceptually subpopulations 'live' on geographically separated regions [203]. A new operator – *migration* – controls the process of exchanging individuals between regions (see Fig. 1.2). The model is usually described by a few parameters: a number of regions, a number of individuals in each region, as well as migration topology, rate/interval and a strategy of choosing individuals to migrate.

Migration topology describes how individuals migrate from one region to another. This often depends on software architecture and the most common are hypercube, ring, or k-clique. In an *island* model individuals can migrate to any other subpopulation, while in a *stepping stone* model individuals can migrate only to neighbouring region(s). Migration rate and interval denote how many and how often individuals migrate. Of course migration rate should be greater if migration interval is longer. Typically the best individuals are chosen for migration and immigrants replace the worst individuals in a destination region. Other possibility is that immigrants replace the most similar individuals (e.g. using Hamming distance as a measure) or just replace emigrants.

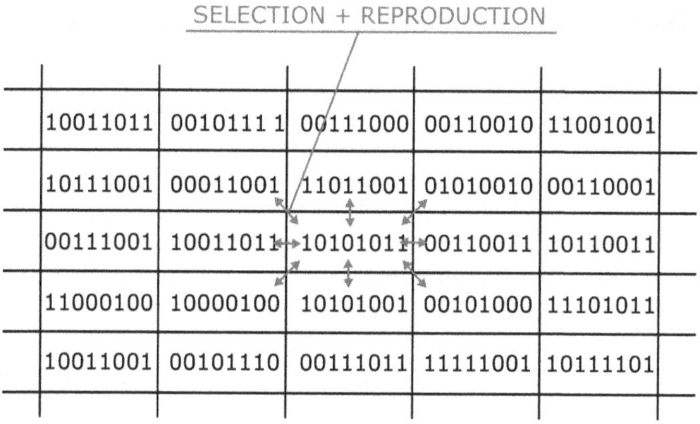

SELECTION + REPRODUCTION

10011011	0010111 1	00111000	00110010	11001001
10111001	00011001	11011001	01010010	00110001
00111001	10011011	10101011	00110011	10110011
11000100	10000100	10101001	00101000	11101011
10011001	00101110	00111011	11111001	10111101

Fig. 1.3 Cellular parallel evolutionary algorithm

Cellular Models

In cellular PEA a population has a fine-grained structure with neighborhood relation defined [232]. In this case local selection/mating is performed over neighbouring individuals — subpopulations are not geographically separated but rather overlap and consist of a few individuals (even only one individual). Usually the neighborhood structure is a grid (each individual has a fixed number of neighbors) with a torus topology. A parameter is needed to describe a neighborhood radius, which defines a direct neighborhood of an individual where selection and mating is performed (see Fig. 1.2).

This model is strongly related to massively parallel computing. In such cases usually each processor contains one individual and is responsible for calculating its fitness and executing genetic operators. Neighbourhood topology is then determined by MPC architecture (Fig. 1.3).

1.3 Sophisticated Population-Based Methods

One of the most important aspects of this monograph are population-based metaheuristics. Besides the classic ones, like evolutionary algorithm, some of more complex methods are also of interest. In particular, hybrid methods such as cultural and memetic algorithms, also immunological algorithms can be integrated with the proposed computing system to yield interesting results.

1.3.1 Hybrid Search Methods

Talbi in [277] provides a concise way of classification of hybrid approaches based on selected design issues:

- In low-level hybrid techniques, a certain function of one algorithm is replaced with another optimization algorithm.
- At the same time, in high-level hybrid techniques, different optimization algorithms are combined without changing their intrinsic behavior.
- In relay hybrid techniques, the individual algorithms are applied in a line, one by one.
- In teamwork hybrid techniques, each algorithm performs an independent search.

Based on the above-mentioned issues, Talbi [277, 278] identifies four main groups of hybrid metaheuristics:

- LRH (low-level relay hybrid): an optimization method is embedded in a single-solution algorithm (e.g. hybridization of local-search method inside a simulated annealing).
- LTH (low-level teamwork hybrid): an optimization method is embedded in a population-based algorithm (e.g. memetic systems hybridizing local-search with evolutionary computation).
- HRH (high-level relay hybrid): several optimization algorithms are executed in a sequence (e.g. local-search yields an initial population for evolutionary algorithm, then the results of evolution are processed by tabu search).
- HTH (high-level teamwork hybrid): several optimization algorithms executed in parallel, cooperate to find a solution (e.g. parallel evolutionary algorithm).

The computing systems discussed in this monograph cross two classes: LTH and HTH.

Again, following Talbi [277], other important attributes of hybrid algorithms may be identified:

- Homogeneous and heterogeneous hybrids combining the same or different algorithms (e.g. parallel evolutionary algorithm consisting of identically configured islands vs. differently parametrized or completely different algorithms running on the islands).
- Global and partial hybrids combining the algorithms searching the whole space or its part (e.g. evolutionary algorithm vs. coevolutionary algorithm).
- General and specialist hybrids solving the same or different target optimization problems (e.g. memetic algorithm vs. meta-evolution [137]).

Following this classification, the systems discussed in this monograph may be perceived as homogeneous, global and general hybrids.

Both single-solution and population-based approaches may be used to find the final solution, however, the latter seem to suit the difficult problems better (i.e., black-box optimization). Population-based approaches produce subsequent approximations of the solution in parallel, utilizing more information about the solutions already found (by the means of variation operators such as crossover and mutation) than single-solution oriented approaches (such as simulated annealing).

To sum up, a combination of two or more metaheuristics can be perceived as a hybrid metaheuristic. This class of algorithms is very simple from a practitioner's point of view (their implementation based on connecting different "components" is easy).

1.3.2 Cultural Algorithms

The cultural algorithms extend the evolutionary computing field by adding new capabilities of influencing the search process: the cultural space and multi-level interactions (not only cooperative and competitive relations imposed by the selection procedure and new population generation, but also the belief or knowledge space, along with the possibilities of its genetic modification is considered). Cultural and memetic algorithms (described in the next section) are inspired by Richard Dawkins' theory of cultural evolution [72], which assumes that culture may also be decomposed into self-replicating parts (so-called memes) that will compete to prevail in the environment.

Culture includes habits, knowledge, beliefs, customs, and morals of members of society. Culture and environment where the society lives are interrelated, and the former interacts with the latter, via positive or negative feedback cycles. Culture also influences the individual evolution, by constructing different rules affecting, e.g. the possibilities of reproducing of certain individuals (compare, e.g. castes in India).

Thus, the cultural algorithm may be defined as a multi-level system that takes advantage of both evolutionary and cultural relations between the individuals in order to influence the ontogeny and manage the search process. For example, during the development of evolutionary search individuals accumulate information about the environment and its different features. This information constitutes the belief space also known as knowledge base and it (or its parts) can be communicated to other individuals. Then, feedback is given, e.g. by communicating interesting areas of the search space (positive feedback) or issuing a warning about the pitfalls (negative feedbacks).

Individuals become units who are capable of learning, being at the same time subjects of some embedded search technique (typically an evolutionary algorithm) and utilizing a higher-order mechanism to process the cultural information (e.g. some inference engine) to affect the evolution search process. The knowledge base may contain:

- Normative knowledge: e.g. a collection of value ranges for individuals, acceptable behavior etc.
- Domain specific knowledge: intrinsic information about the domain of the problem.

- Situational knowledge: information about the ongoing search events (e.g. the best solution so far).
- Temporal knowledge: history of the search space, e.g. a tabu list.
- Spatial knowledge: information about the topography of the search space.

The general algorithm of a cultural metaheuristic is presented in Pseudocode 1.3.1. One of an interesting uses of this approach is Learnable Evolution Model proposed by Ryszard Michalski [213], utilizing an inference engine to guide the evolutionary search process. Cultural algorithms may be perceived as a hybridization of evolutionary, or generally, population-based metaheuristic, with a predefined knowledge base used as a reference point by certain operators, though it does not fit into the Talbi's classification (see Sect. 1.3.1).

Pseudocode 1.3.1: PSEUDOCODE OF CULTURAL ALGORITHM

population ← *initialisePopulation*()
knowledgeBase ← *initialiseKnowledgeBase*(*population*)
while not *stoppingCondition*()
⎰ *evaluations* ← *evaluate*(*population*)
⎪ *matingPool* ← *selection*(*population*, *evaluations*)
⎪ *population* ← *crossover*(*matingPool*)
⎨ *population* ← *mutation*(*population*)
⎪ *population* ← *influence*(*population*, *knowledgeBase*)
⎱ *knowledgeBase* ← *update*(*population*)

1.3.3   Memetic Algorithms

Memetic algorithms belong to a class of cultural algorithms and historically are evolutionary algorithms enhanced by hybridization with local-search methods (the first successful approach was made by Pablo Moscato [218], who hybridized the evolutionary search with a local improvement, using of simulated annealing to solve Traveling Salesman Problem). The evolutionary algorithm utilizes the local-search method (in the simplest case, the greedy local search or more sophisticated local search techniques, such as simulated annealing or tabu search) within its evolutionary cycle (in the course of evaluation or mutation). A scheme of the memetic algorithm is shown in Pseudocode 1.3.2. Note that usually only one memetic local search procedure (Baldwinian or Lamarckian) is used:

- Baldwinian local search according to Baldwin theory predispositions may be inherited during reproduction [16]. This method is usually implemented as local search procedure called in the course of the evaluation process. The evaluated individual is assigned the fitness function value computed for one of its possible descendants (effects of the local-search starting from this individual). First approaches that may be classified as Baldwinian memetic were oriented on so-called Baldwin-effect

[8, 144]. The attained improvement does not change the genetic structure (genotype) of the individual that is transferred to the next generation. The individual is kept the same as before local search, but the selection is based on the improved fitness after local search. Baldwin effect follows natural evolution (Darwinian), i.e., learning improves the fitness and selection is based on fitness. The improvement, in this case, is passed indirectly to the next generation through fitness.

Pseudocode 1.3.2: GENERAL PSEUDOCODE OF MEMETIC ALGORITHM

population ← *initialisePopulation*()
while not *stoppingCondition*()
⎡ *evaluations* ← *BALDWINIANEVALUATION*(*population*)
⎢ *matingPool* ← *selection*(*population, evaluations*)
⎢ *population* ← *crossover*(*matingPool*)
⎣ *population* ← *LAMARCKIANMUTATION*(*population*)

function *BALDWINIANEVALUATION*(*population*)
return *localSearchFitnesses*(*population*)

function *LAMARCKIANMUTATION*(*population*)
return *localSearchGenotypes*(*population*)

- Lamarckian local search according to Lamarck's theory characteristics of individuals acquired in the course of life may be inherited by their descendants [105]. This method is usually implemented as a local search procedure called in the course of execution of mutation or crossover operator. The search for a mutated individual is based not only on a stochastic one-time sampling from the solution space, it may be a much more complex process, being an outcome of the local search starting from this individual. In the same way the memetic crossover may be implemented, by trying different combinations of parents' genotypes, until a satisfactory is found. In Lamarckian evolution, individuals improve during their lifetime through local search and the improvement is passed to the next generation. The individuals are selected based on improved fitness and are transferred to the next generation with the improvement incorporated in the genotype.

Although both theories have not been fully verified, metaheuristics based on them are effective in many problems (see, e.g. [187, 294]). Memetic algorithms retain exploration properties of evolutionary algorithms with enhanced exploitation based on the local-search. This makes memetic algorithms faster, but the problem of diversity loss arises and must be duly dealt with [222].

In a comparison of Baldwin and Lamarckian learning, Whitley et al. [294] showed that utilizing either form of learning would be more effective than the classical genetic algorithm without any local improvement procedure. Though Lamarckian learning is faster, it may be susceptible to premature convergence to a local optimum as compared to Baldwin learning. Yao [303] examined both Lamarckian evolution and Baldwin effect in combination with an evolutionary algorithm and local search. The obtained results show that there is no significant difference between Lamarckian-evolution-style combination and Baldwin-effect-style combination.

When implementing a memetic algorithm, it must be decided where (in which phase of evolutionary computation) and when (in which evolutionary cycle) the local search should be applied. Moreover, it must also be decided which individuals need to be improved, how long the local search should run and which memetic model should be used (Baldwinian or Lamarckian)? It is noteworthy that this hybridization introduces additional complexity to the evolutionary algorithms, emphasizing their applicability as the last resort methods (cf. [211]).

Krasnogor and Smith [183] provide a formalization of the means of hybridization of evolutionary algorithms with local search methods by introducing a concept of schedulers. These schedulers are used to decide when and where to introduce the local search during the evolution process:

- Fine-grained mutation and crossover schedulers can replace the mutation and crossover operators running the local-search (in this way, e.g. Lamarckian memetics are implemented).
- Coarse-grained scheduler can replace the selection (in this way, e.g. Baldwinian memetics are implemented) and may decide which local search methods are used, which individuals are subject to this process etc.
- Meta-scheduler retains and provides historical information to the local search method, making possible implementation of tabu-search like techniques.

The memetic algorithm may consist of all schedulers, but whether it becomes better, or simply more complex, can only may be discovered in the course of experiments.

Memetic algorithms may be perceived as LTH hybrids, according to Talbi's classification (see Sect. 1.3.1), merging population-based (evolutionary) system along with a local-search technique.

1.3.4 Immunological Techniques

Immunology, which a branch of biomedical science, is situated between medicine and biology, is over 100 years old. Louis Pasteur and Robert Koch are said to have been its founders, as they were the first to carry-out research on disease-causing micro-organisms.

The notion of "artificial immune systems" refers to a class of systems based on ideas, theories and components inspired by the structure and functioning of immune system of vertebrates [74]. Immunological algorithms form an interesting family of metaheuristics inspired by the work of immune system of vertebrates. These techniques are younger than evolutionary algorithms, however they have already found a vast number of applications e.g. for anomaly detection, classification, optimization and many others [295].

The most important notions used in immunology, both biological and artificial one are *pathogens* or *antigens* (intruders that should be removed from the system), *lymphocytes* or *antibodies* (immune cells that should clean the organism from the pathogens) and *affinity* (treated as a measure of similarity of the immune cell to the antigen). Working of the immune system consists in construction of lymphocytes containing patterns of the foreign molecules. A correctly working immune system should be able to remove the foreign microorganisms, thanks to the similarity (affinity) between the lymphocytes and the foreign cells, at the same time tolerating the self cells (belonging to the organism).

The beginning of artificial immune systems dates back to the 1980s, when a proposal was made to apply theoretical immunological models to machine learning and automated problem solving [106, 146, 272]. These algorithms received more attention at the beginning of 1990s (see, e.g. [23, 152]).

First research studies in the field were inspired by theoretical models (e.g. immune network theory) and applied to machine learning, control and optimization problems. Computer Immunity by Forrest et al. [115, 118] and Immune Anti-Virus by Kephart [165, 166] are classical examples. These works were formative for the field as they provided an intuitive application domain captivating a broader audience and assisting in differentiating the work as an independent sub-field.

Modern Artificial Immune systems are inspired by one of three sub-fields: clonal selection, negative selection and immune network algorithms. The techniques are used for clustering, pattern recognition, classification, optimization, and other similar machine learning problem domains [73]. Below, negative selection and clonal selection algorithms will be discussed in detail, as they provided inspiration for part of the research presented later in this monograph.

Generally, artificial immune system functions as a classifier, identifying certain patterns (similar to detecting intruders—antigens in the immune system of vertebrates). Therefore, dedicated heuristics to solve classification problems (based on creating antibodies identifying certain patterns) can be easily implemented. However, search and optimization problems may also be solved by an immune approach (after undertaking certain assumptions, e.g. treating the optima of the goal function as antigens that should be identified) [69].

There exist a number of algorithms that map immunological inspirations directly to numerical optimization. They may be treated surely as an interesting alternative for evolutionary algorithms. Because they are also population-based techniques, they are often called *immunogenetic algorithms* [295]. An example of such algorithms is *constrained genetic search*, that was developed for searching for function optimum while maintaining a number of constraints [140]. In this approach the feasible solutions are treated as antigens and infeasible as antibodies. The antigen set consists of the solutions with the highest fitness function value. Based on the assumption, that antigens will be able to detect substantial antigens, it can be assumed, that thanks to modification of these antigens, one can lead to "repairing" of their genotypes and obtaining less-infeasible solutions. Another example is a clonal selection algorithm modified in order to solve multi-modal function optimization. The modification consists in treating the unknown function extrema as antigen set. Affinity of the antibody to the antigen becomes the value of the goal function [75].

The main two immunological mechanisms used during construction of the antibodies pool are clonal selection and negative selection [73, 74].

Clonal Selection

The clonal selection process serves as an inspiration for a group of algorithms that may be applied to classification or optimization problems. The clonal selection theory was proposed by Burnet to describe the behavior and capabilities of antibodies in acquired immunity [38]. According to this theory, the presence of antigens causes a selection of the most similar lymphocytes. When the lymphocyte is selected, it binds with the antigen and proliferates, creating thousands of copies of itself (affected by small copying errors so-called somatic hyper-mutation, broadening the possibilities of detecting the antigens). The lymphocyte creates two main types of cells: short living plasma cells that help to remove the antigens by producing antibody molecules, and long-lived memory cells active in a secondary immune response when the antigen is found again in the organism.

Thus this strategy may be seen as a general learning method, including a population of adaptive information units (representing sample solutions to the problem) subject to a competitive process based on selection. This selection consists in proliferation with mutation and produces a generation of individuals better that become fitter to solving the problem.

The general clonal selection algorithm (see Pseudocode 1.3.3) involves a selection of antibodies (candidate solutions) based on affinity, computed either by matching against an antigen pattern (in classification problem solving) or via evaluation of the pattern by the cost function (in optimization problems). Selected antibodies are subject to cloning, and the number of cloned antibodies is proportional to affinity. At the same time, the hyper-mutation of clones is inversely proportional to clone affinity. The resulting set of clones competes with the existing antibody population for membership in the next generation. In the end, several low-affinity members of the population are replaced with randomly generated antibodies. When the classification problem is considered, the memory solution set must be maintained in order to represent the solution patterns.

Other algorithms based on clonal selection that are worth mentioning are: CLIGA algorithm [67], the B cell algorithm [162], and Forrest's algorithm [116].

Negative Selection

The negative selection algorithm was inspired by "self/non-self discrimination" observed in the acquired immune system of vertebrates. The proliferation of lymphocytes during the clonal selection makes possible to generate a wide range of detectors to cleanse the organism of harmful antigens. What is in this process, is that self-cells belonging to the organism are avoided and should not be detected by the lymphocytes in the case of a healthy immune system. In other words, during the process, no self-reactive immune cells are created. Such a set of antibodies is

Pseudocode 1.3.3: PSEUDOCODE OF CLONAL SELECTION ALGORITHM

$population \leftarrow initialisePopulation()$
while not $stoppingCondition()$
$\begin{cases} affinities \leftarrow computeAffinities(population) \\ selected \leftarrow selection(population, affinities) \\ clones \leftarrow clone(selected, affinities) \\ clones \leftarrow hypermutate(clones) \\ clonesAffinities \leftarrow computeAffinities(clones) \\ population \leftarrow population \cup chooseBest(clones, clonesAffinities) \\ population \leftarrow population \setminus chooseWorst(population, affinities) \\ population \leftarrow partialRandomReplace(population) \end{cases}$

achieved during proliferation, due to a negative selection process that selects and removes the autoimmune cells (that bind with self-cells). This process is observed in nature during the generation of T-lymphocytes in thymus.

The self/non-self discrimination process, which uses a negative selection, consists in creating the anomaly and change detection system modeling anticipation and variation based on a certain set of well-known patterns. Thus, an appropriate model is built by generating patterns that do not match an existing set of available (self or normal) patterns. The resulting non-normal model is then used to detect the unknown by matching the newly-received data to the non-normal patterns.

The principles of this algorithm are quite simple (see Pseudocode 1.3.4). After generating random detectors, they are matched against the set of self-patterns. Those matching ones are removed, while non-matching are kept in the detector repertoire. When the repertoire is ready (a certain number of detectors has been reached), then, based on the set of non-self immune cells, a classification of foreign patterns may be performed (again, according to the same affinity measure that was used to generate this set of antibodies).

Pseudocode 1.3.4: PSEUDOCODE OF NEGATIVE SELECTION ALGORITHM

$self \leftarrow set\ of\ self\ patterns$
$repertoire \leftarrow \emptyset$
while not $isComplete(repertoire)$
  do $detectors \leftarrow generateRandomDetectors()$

$selfClass \leftarrow \emptyset$
$nonSelfClass \leftarrow \emptyset$
for all $d \in detectors$
$\begin{cases} \textbf{if not } match(d, repertoire) \\ \quad \textbf{then } selfClass \leftarrow selfClass \cup \{d\} \\ \quad \textbf{else } nonSelfClass \leftarrow nonSelfClass \cup \{d\} \end{cases}$

The first negative selection algorithm was proposed by Forrest [118] and applied to the monitoring of changes in the file system (corruptions and infections by computer viruses), and later formalized as a change detection algorithm [80, 81].

1.4 Summary

Selected metaheuristics that were presented in this chapter do not exhaust, however, a plethora of such methods that are available nowadays (cf., e.g. [278]). The important thing is that these methods are often hybridised, and the resulting systems become a valuable weapons of choice when dealing with complex "black-box" problems. Among popular approaches, hybrid systems utilizing tabu search with other metaheuristics (see, e.g. [279, 288, 289]), or simulated annealing (see, e.g. [188, 189, 305]) yielding successful results may be considered. Therefore, further search for hybrid systems is worth undertaking because looking for synergy of different approaches may lead to obtaining efficient methods in solving particular difficult problems.

In particular, agent-oriented hybrids seem to have much potential, as the feature of autonomy, makes it possible to employ various techniques not exploited before. Hybridisation with different local search methods is of course possible (including tabu search, simulated annealing, or different memetic-oriented approaches). Moreover, particular agents may decide whether to use or not a certain local search operator or any other technique, to adapt parameters of their search operators in order to maintain a balance between exploration and exploitation of the search space.

Designing complex search methods needs not only employing sufficient methods for validation of their outcome, but also verification of their design and analytical features.

Several formal models aimed at proving different features of evolutionary metaheuristics have been constructed. One of the first and important models of metaheuristic methods was Michael Vose's SGA model [290], which proves that simple genetic algorithm (SGA) with the population of a fixed size [133] can be modelled by a Markov chain, and after further assumption that the mutation rate is positive, this chain is ergodic. This result, which formally confirms that SGA is a well-defined global optimization algorithm (belonging to Las Vegas class [149]), become an inspiration for preparing a Markov-chain based models for agent-based metaheuristics (see Chap. 3) and proving their ergodicity feature.

Other approaches to model evolutionary algorithms, which are worth mentioning, are different models for single-population, e.g. [71, 134, 197, 246–248, 274] and parallel evolutionary algorithms, e.g. [3, 109, 280]). Using some of these models, the researchers (e.g. [249]) have proved selected practical features such as first hitting time (see, e.g. [25]) for simple settings (e.g. evolution strategy $1 + 1$ solving the problem of optimization of a convex function).

There also exist other extensions of Vose's model dealing with hybrid metaheuristics (see, e.g. [50, 255]). Though the formal proofs of asymptotic features are unavailable with one extension, the ergodicity feature has been proved by Schaefer et al. for parallel Simple Genetic Algorithm in [256], as an extension of the works by Vose.

Apart from these systems, more advanced search and optimization techniques like memetic or agent-based computational systems lack such models, with some notable exceptions. For example, [302] considers an abstract model of memetic algorithms based on the application of gradient-based local search to the whole population on a generation-basis and provides a sufficient condition for quasi-convergence (i.e., asymptotically finding one of the best k solutions in the search space, where k is the population size). Also, [227] considers adaptive MAs and indicate that only static, greedy and global adaptation strategies (i.e., strategies that use no feedback, check all possible memes and pick the best one, or use a complete historical knowledge to decide on the choice of meme respectively) are globally convergent using elitist selection mechanisms. This stems from [56, 245] and cannot be proved in general for local adaptive strategies.

In general, there is still a lack of a comprehensive stochastic model of the wide class of population-based, agent-based, cultural or memetic computing systems. It may result from the fact that such systems are still poorly defined and formalised, which of course has to be done before making any attempt to analyze their features analytically. Moreover, these systems are complex, so thorough preparations for detailed formal model leading to significant theoretical results are complicated and very time-consuming.

It is easy to see that careful preparation of general models for such systems will surely lead to a better understanding of them and it is necessary to make sure that one does not only propose a complex metaheuristic, but provides a potential user with a firm background, proving that it makes sense to use such a complex computing method. At the same time, above-mentioned models of metaheuristics (cf. Rudolph [249]), focused on proving certain features (e.g. first hitting time), considering even simple, convex functions, should not be underestimated. Again, referring to the well-known "no free lunch theorem", it is hardly probable that one model can fulfil all the researchers' needs. Therefore various approaches to modeling of metaheuristics, offering results of different types will always be valuable.

Chapter 2
Agent-Based Computing

In the 1970s, there was a growing interest in the systems, where a task to be solved was decomposed into smaller parts (subtasks), in order to solve them separately and later integrate the solution. This approach may be described as *distributed problem solving*, and it is usually easy to implement in parallel environments such as multi-core machines, clusters or grids. In the field of *multi-agent systems* bearing a significant legacy from the distributed problem solving, distributed individuals (agents) received great attention, mainly because they were perceived as autonomous beings, being capable of interacting with their environment and other agents, bearing the features of intelligence.

In fact, during the last decades intelligent and autonomous software agents have been widely applied in various domains, such as power systems management [207], flood forecasting [128], business process management [153], intersection management [89], or solving difficult optimization problems [191], just to mention a few. The key to understand the concept of a multi-agent system (MAS) is intelligent interaction (like coordination, cooperation, or negotiation). Thus, multi-agent systems are ideally suited for representing problems that have many solving methods, involve many perspectives, and/or may be solved by many entities [299]. That is why, one of important application areas of multi-agent systems is large-scale computing [32, 286].

This chapter sketches the basic definitions of agent and agent system, then describes popular agent-based architectures particularly useful for computing systems. Next, evolutionary agent-based computing paradigm is presented along with detailed description of Evolutionary Multi-Agent System (EMAS) along with some of its variations.

© Springer International Publishing AG 2017
A. Byrski and M. Kisiel-Dorohinicki, *Evolutionary Multi-Agent Systems*,
Studies in Computational Intelligence 680, DOI 10.1007/978-3-319-51388-1_2

2.1 Agency, Architectures, Management and Computing

Beginning of agency can be dated back to 50s of 20th century. One of the first scientists working on agents were John McCarthy and Oliver Selfridge, who proposed the notion of "agent" for the first time [161]. Agent-based techniques cannot be perceived as one, homogeneous methodology, as they originate from many different concepts and applications. Some of them develop quicker and more intensively (e.g. interface agents used for facilitating everyday computer use), other are still researched and will be evaluated properly in the future. Advantages of agent-based systems can be seen in many applications (e.g. simulation and control of transportation systems). Research on other application areas (e.g. agent-based computing) is going on.

Starting in the half of 70s of the 20th century, many scientists working on artificial intelligence began to appreciate the systems using the decomposition of the task to many smaller ones, in order to solve them one by one and later combine them into the global solution of the given problem. The approach called *distributed problem solving* focused on the possibilities of decomposing the computing process into smaller parts [103]. In the multi-agent systems a greater attention was given to the particular units of distribution, that were endowed with autonomy and could perceive their environment in order to achieve its goals.

Agents play an important role in the integration of artificial intelligence subdisciplines, which is often related to a hybrid design of modern intelligent systems [250]. In most similar applications reported in the literature (see, e.g. [62, 252] for a review), evolutionary algorithm is used by an agent to aid realization of some of its tasks, often connected with learning or reasoning, or to support coordination of some group (team) activity. In other approaches, agents constitute a management infrastructure for a distributed realization of an evolutionary algorithm [284–286].

2.1.1 Agents and Multi-agent Systems

According to one of the most popular definitions proposed by Wooldridge, an agent is a computer system situated in an environment, and capable of undertaking independent, autonomous actions in this environment in order to fulfill tasks on behalf of its user [300]. Autonomy is perceived as one of the most crucial features of an agent.

Following this definition, any computer program that manages a certain apparatus (e.g. a thermostat) or affects the state of the computer system (e.g. a Unix system daemon) may be perceived as an agent. Thus it might seem that such definition of an agent introduces a new name for some existing, well-known programming techniques. In fact, intelligent agents bring new quality crossing the borders of already existing computer systems, enhancing the notion of an object or process with additional, important features, e.g. [79, 300] helping other agents to fulfil their goals:

- reactivity: agents may perceive their environment and react to changes in that environment,
- pro-activity: agents may perform tasks based on their own initiative,
- social ability: agents are able to interact with other agents (also with users).

Another classification of agents should be referred to, in order to see how autonomous agents undertake their decisions [300]:

- logic-driven agent—utilizing deduction process [126],
- reactive agents—undertaking the decisions uses a predefined function mapping the situations into actions [35, 107],
- BDI (*ang. belief-desire-intention*) agents—manipulate the data structures representing beliefs, goals and intentions of the agents [34, 59, 60].

It is noteworthy that fulfilling the goal becomes a *raison d'être* for an agent. This is also the most important and determining factor in undertaking actions in the environment by an agent.

The notion of agent system is based directly on the notion of agent. Generally speaking, an agent system is a system, in which a key abstraction is that of an agent. Therefore, a multi-agent system is one that consists of a group of agents which interact one with another [108, 156].

Agents act in their environment, and different groups of agents may perform their tasks in different parts of the environment. In particular, their activities may overlap. As an example, the possibility of communicating between agents that are "close" in the environment may be given (of course, their closeness depends strongly on the notion of neighborhood, if it was implemented), or direct interaction with the environment (e.g. only one agent-robot may pass through the door at a time) [299].

Among the main features of multi-agent systems one may distinguish [79]:

- distribution—the agent system structure is easy to implement in distributed systems (e.g. in a local cluster),
- autonomy—each agent exists independently on the others, it can sustain in the environment without the external interaction, it undertakes the decisions based on its observations, desires and its own model of the environment,
- decentralization—there should be no global control mechanisms for the whole agent system as a consequence of the autonomy of the particular agents,
- exchange of beliefs—agents can communicate in order to exchange their information about the environment, their actions etc. (using e.g. ontologies to describe their world),
- interaction—agents can interact with other agents and with the environment by exchanging the messages using the predefined protocols or standard communication languages as e.g. ACL (Agent Communication Language) or KQML (Knowledge Query Manipulation Language) [110, 127],
- organization—decentralization and autonomy in the distributed environment creates the need to introduce a certain order of the agent population, moreover the agent usually can perceive only part of its environment and part of the neigboring agents,

- situatedness—the environment naturally sets constraints for all the possible actions and observations undertaken by the agent,
- openness—it is often difficult to set the structure of the environment that is heterogeneous and can be dynamically changed (e.g. Internet network),
- emergence—arising of holistic phenomena not programmed before, usually shown by a certain group of agents (i.e. collective intelligence) [135, 136],
- adaptation—these systems are flexible and can adapt to the changing environmental conditions,
- delegation—an important notion, following Bradshaw: "agent is that agent does",
- personalization—the agent may acquire the preferences of its user in order to ease his/hers central activities (e.g. interface agents facilitating typewriting) [195],
- accessibility—the agent systems are quite easy for description, application and adaptation to many problems, they are also very easy for implementation, providing that a certain framework is available, supporting different low-level services such as communication or monitoring.

Agents have also been used in computing systems to enhance the search and optimization capabilities with the above-mentioned agency features.

2.1.2 Architectures of Agent-Based Computing Systems

Starting with a multi-agent perspective, three types of computing agent system architectures, which form consecutive levels of increasing complexity, can be distinguished [177]. The main purpose of these architectures is to serve as a means of incorporation of computing methods into cooperating, autonomous entities—agents.

Figure 2.1 presents an example of a classical approach to hybrid computing system—an evolutionary system using a stochastic hill-climbing algorithm constituting in fact a memetic system. Each individual represents a solution, and the evolving population explores and exploits the feasible search space. The whole system is embodied in an agent utilizing this idea to adapt to the specific environment.

Evolutionary multi-agent system shown in Fig. 2.2 can be seen as next step in possible specialization [61]. In this case, evolutionary processes work at a population level—agents searching for a solution to a certain problem are able to generate new agents and may be eliminated from the system on the basis of adequately combined evolutionary operators. The predefined distributed selection mechanism increases the possibility of reproduction for the best agents (in terms of fitness function value). The result of the search is formed as a set of results obtained by single agents. The architecture of EMAS is homogeneous, which means that the agents are identical as far as an algorithm and built-in actions are considered.

EMAS may be easily hybridized with local search techniques forming a memetic variant of the base system. In this case, each agent performs local search in the course of its life in the system. This may be carried out during reproduction or at arbitrarily chosen moments.

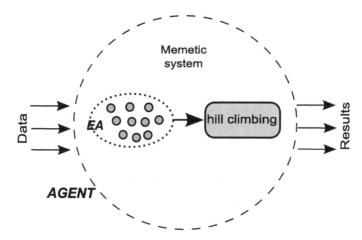

Fig. 2.1 A hybrid computing system located in a single agent

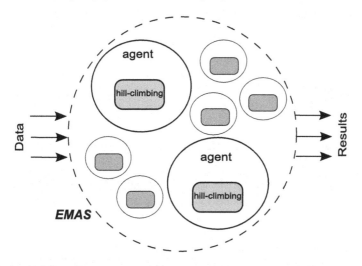

Fig. 2.2 A population of homogeneous agents

The highest specialization level is reached when the agent population is heterogeneous (see Fig. 2.3). There exist in the system different agents, focused on solving different problems or realizing of different task related to problem solving or system organization, communication etc. The result of the system is stipulated as a consequence of the outcome of the negotiation process among the agents. Many different techniques and protocols can be applied to this purpose. It may be said that these systems closely approach typical multi-agent ones operating in the computer network environment [177].

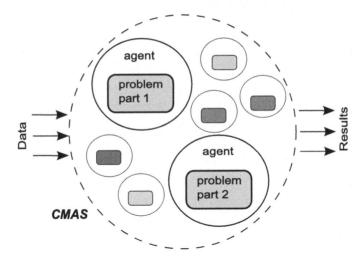

Fig. 2.3 A population of heterogeneous agents

EMAS (Sect. 2.3) and its memetic variant (Sect. 2.4.1) belong to the second class, while other variants of EMAS (Sect. 2.4) described in this monograph belong to the third class described above.

2.1.3 Computing Systems Management by Agent Means

One of the possible applications of agency in computing systems is situated at a technical level. An agent community may take care of the the management of the distributed system, utilizing their autonomy and auto-adaptation to implement, e.g. scheduling or load balancing.

Distributed management of the system deployed in a cluster or grid requires a definition of node topology, neighborhood and migration policies that are affected by a current load of nodes. Well-known standards for constructing multi-agent environments (such as FIPA [111]) do not provide such capabilities. Another important functionality missing in many platforms is the notion of distance between agents, measured as a factor of communication throughput.

Grochowski and Schaefer [284–286] proposed Smart Solid Architecture (see Fig. 2.4) that supports these requirements. This architecture is similar to EMAS architecture (cf. Sect. 2.3), where agents are also homogeneous.

However, in this approach the task is divided into subtasks and these are delegated to agents which are to solve them. Agents in this model accomplish two goals: perform computation of the task and try to find a better execution environment (computing node) for the task, based on the load and throughput information. It is

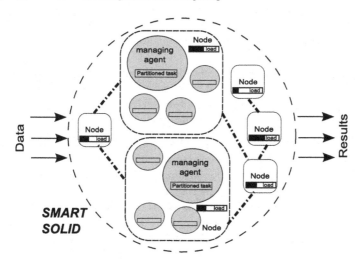

Fig. 2.4 Smart solid architecture

noteworthy that agents do interact with one another, in order to be able to solve the assigned task. Agents may undertake the following actions:

- execute the task in order to solve it and communicate the results to other agents,
- denominate the load requirements,
- based on the requirements compute the possible migration capabilities of the agent,
- further divide the task, create the child agents,
- migrate to another execution environment carrying the task.

An effective scheduling algorithm was also introduced in the discussed environment and tested by Schaefer, Grochowski and Uhruski, utilizing the natural inspiration of the diffusion mechanism [138]. This system was successfully applied e.g. in the problem of mesh generation for computed-aided engineering applications [257].

2.2 Bringing Together Agent-Based and Evolutionary Computing

Recalling the most important features of multi-agent systems (MAS) that are adopted by the researchers [108] and can be regarded as definitions of the term:

- a multi agent system is constituted by a set of agents and a sub-system called environment;
- agents perform certain actions that are characteristic to them, resulting in changes of themselves or the environment;

- some of elementary agents' actions form mechanisms perceived as specific in multi-agent systems in general, e.g. negotiation, communication, exchange of resources etc.;
- providing that there is a structure embedded in the environment (a graph) reflecting spatial phenomena of the system, it forms a base for migration of agents;
- an agent that is treated as a black box (observed from the outside) may have some human attributes like autonomy, intelligence etc.

one can consider building an agent-based system, focused on solving computing tasks, incorporating the above-mentioned features. However, one must keep in mind, that computing systems are aimed at producing tangible results (e.g. an optimal value of certain function, an optimal set of certain parameters), thus they need certain mechanisms for synchronization of their work. At the same time any synchronization realized in agent-based systems is suspicious, as it may be perceived as a means for global control. Agent-based systems leverage the notions of autonomy, thus bringing up new techniques for control of the agents that are aimed at producing one common goal are necessary.

2.2.1 Non-renewable Resources

The most important problem of the outlined concept is that in agent systems using centralized mechanisms of computing process is impossible. A specific approach must be created, because of:

- lack of global knowledge and global synchronization of actions,
- autonomy of particular agents in undertaking their actions.

Thus the concept of controlling the evolution process based on non-renewable resources was proposed [169, 174].

The developed mechanisms demand defining a certain number of resources, that describe the state of the environment and agents. The resources can be used to indicate the quality or efficiency of the agents solving particular task. The agents rich in resources may perform certain actions and survive in the environment. It is usually realized by defining of a certain threshold of the resource, required for realizing particular action. The algorithms of the agent behavior may also be defined by the probability distributions dependent on the amount of resources owned. To very important issues belong the boundary conditions of the of the agent resources, conditioning its participation in the processes realized in the population, thus affecting the behavior of the whole system. On exceeding these conditions the agent is removed from the population and its remaining resources are returned to the environment or distributed among the other agents.

Each agent during creation in the system obtains certain amounts of the resources, either from other agents or from the environment. Realization of certain actions may make the agent lose or gain resources. This can become a mechanism for assessing the quality of the agent—increasing of the resource amount may be treated as a reward for "good" behavior, and decreasing as a punishment. The most natural way of realization of this mechanism is introducing a cost for realization of any action and rewards for achieving good results of the realized task. The resources may be assigned using the environmental mechanisms on the local level or by the dedicated agents based on information about the quality of the reached solutions, supplied from outside.

2.2.2 Integrating Evolutionary Paradigm with Agency

A starting point for the idea of *agent-based evolutionary computing* was the observation that many newly constructed problem classes calls requires new metaheuristics, in particular evolutionary approaches, on the other hand classic evolutionary computing techniques do not follow many essential aspects of natural evolutionary processes [14]. The basic evolutionary algorithm is often enhanced in order to realize specific mechanisms observed in nature, in particular in order to maintain useful diversity of the population. At the same time, one of the most neglected fact of classic evolutionary computing is that the whole evolution process is centralized and fully synchronized. Thus the objects of evolution become simple, inactive data structures, and they are processed by a common "manager" which has complete and total "knowledge" about the whole population [58].

New possibilities should be given by an evolution model, that will enrich the individuals in capabilities of autonomous acting and interaction in a common environment. As certain substitutes of such approach may be treated several coevolutionary models or parallel evolutionary algorithms, in particular their hybrid variants, making the realization of the evolutionary processes dependent on the particular context of the individual or of the (sub-)population. It seems, however, that using the concept of agency may bring new quality into the computing systems by creating a new computing model. Thus, embedding the evolutionary processes into agent system will make possible full decentralization of the evolution process. Independence of the environment perception and the actions undertaken by the agents makes possible their mutual interaction, leading to opening new possibilities of introducing completely new elements into the evolution process. Such an approach, that may be called *decentralized evolutionary computation* should help in eliminating some of the above-mentioned limitations of the classic evolution model. Similar assumptions are shared by a concept of Multi-Agent Genetic Algorithm [304], however this approach assumes a very limited autonomy of the agents localized in predefined, constant positions of two-dimensional grid (similarly to cellular model of evolution described in Sect. 1.2.4).

A key concept of agent-based evolutionary computing is introducing of evolutionary mechanisms into agent environment, in other words subjecting the agent population to the evolution process. As a result, a system capable of self-adaptation to the changing conditions both on the particular components level (agents) or their organizational structure [84, 221]. It can be said, that the evolution process is "managed" or "lead" to a certain extent by the population of autonomous agents, representing individuals that undergo the evolution process. These individuals may belong to different species or groups.

2.2.3 Logical and Physical System Structure

A very important feature for the evolution process conducted in the agent system is the fact of co-existence of the agents in a common, possibly spatial environment. The agents and resources can be localized in certain places of the environment (regions, cells) where they can be observed, and affected by the other agents (cf. Fig. 2.5). Each individual may access only a certain part of the information and resources, lying in its range. Moreover, the agent can interact, observe and communicate with a limited number of other agents, found in its neighborhood (determined using a predefined topology). An effect of such local perception and local interaction is limited access of the agents to certain resources and information that can be found in the system, thus the evolutionary processes are conducted with higher diversity, making possible implementation of parapatric or allopatric speciation.

Limitations in reachability of resources and information may result in creating somehow privileged areas in the system space (objectively or subjectively dependent on the current state and/or agent needs). This may be also caused by the structure of the system space itself, but also presence (or absence) of agents with certain features. The agents themselves may also want to change their placement in the system, realizing the migration according to the rules imposed by the system space structure. Possibility of migration may also be used for realization of load balancing mechanisms, in the case of the distributed computing environment.

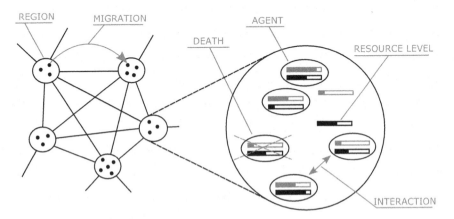

Fig. 2.5 Concept of space and resources in agent-based computing

Pursuing analogies to parallel evolutionary algorithms, one can consider two space models: fine-grained (multi-dimensional neighborhood structure, so-called "cells" with a predefined metric, e.g. "Manhattan" setting the interaction range) or coarse-grained (so-called "regions" or "islands", completely isolated, where the interactions are possible only within one region). In the latter case, the migration range may be limited by the topology of the connections among the regions, similarly to multi-population evolutionary algorithms. Here coarse-grained or hybrid (coarse grained with fine grained inside the regions) models are preferred because of relatively easy implementation and natural mapping of the logical and physical spaces of the agent system.

2.2.4 Representation of the Tasks and Solutions

In the discussed concept, the population of agents is located in a common environment which has certain properties (e.g. space structure, availability of certain resources, placement of agents and resources, possibilities of cooperation etc.) which represent the current state of the real world or certain task to be solved. A significant difference may be perceived here when comparing to classic evolutionary algorithm, where the task is represented only by the fitness function, while the environment of particular individuals is not considered in any way.

Of course agents may observe the environment and the other agents, they may also realize actions affecting their own state, as well as the state of the environment or even other agents. In any possible moment, the state of the agent (e.g. its resources or information), its location in the system space or the outcome of its action (e.g. placement of the resources in the environment) may become a basis for setting its solution of the given problem. The solution may also be represented by the state of the behavior of a group of agents that makes also a significant improvement, when comparing to classic evolutionary algorithms.

The basic variant of the considered model is so-called **evolutionary multi-agent system—EMAS** where the agent is considered as a basic evolution entity [169]. The agent is equipped in the ability of reproduction, that may be accompanied by random changes in the inherited features of the generated agent (mutation, recombination). Moreover, it is necessary to introduce certain mechanisms for elimination of the agents representing low quality solutions. In effect, an asynchronous and decentralized process of evolution is constituted. If it is properly parametrized, it should lead to autonomous adaptation of the system to the state needed in particular situation (i.e. a proper configuration of the parameters).

An interesting alternative to the above-described models is **flock-based multi-agent system—FMAS** [173], where the agent represents a group of individuals (a flock). Similarly to *migrational* model of evolutionary algorithm, in the case of each flock an evolution process is realized (e.g. classic evolutionary algorithm), enhanced by the possibility of migration of individuals. However in this case, the individuals migrate between the flocks within one evolutionary island, while the flocks, that can be treated as another level of system organization, may migrate themselves between

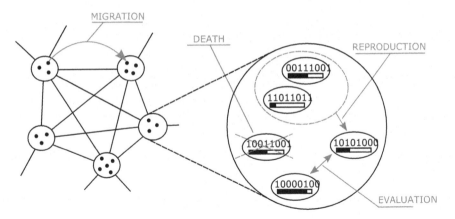

Fig. 2.6 A concept of evolutionary multi-agent system (EMAS)

the islands. Moreover it is possible to split or merge the flocks depending on their state (or rather the state of the individuals constituting the flock), leading to possibility of dynamic reconfiguration of the whole system in the course of the computing process. The presented concept has all the advantages of the parallel evolutionary algorithm, at the same time allowing for dynamic adaptation of the computing system to the problem being solved [172].

2.3 Evolutionary Multi-agent System

The core idea of the evolutionary multi-agent system is treating each agent as an individual in the population, thus equipping it in the information (in the genetic model of evolution—a genotype) allowing for generation (*ontogenesis*) the solution of the given task (*a phenotype*). Generally speaking, the work of EMAS is based on processing the population of agents cooperating in the common environment, according to the inheritance and natural selection rules. However because of the specific features of the agent environment, the evolution must be realized in a different way than in the classic evolutionary algorithm. The key concept of this process will be of course, as it is in the natural processes, the phenomenon of death and the capability of reproduction, that must be reflected in the design of the agents (see Fig. 2.6).

2.3.1 Reproduction and Inheritance

One of the two of the most crucial elements of the evolution process is reproduction, i.e. introducing new offspring individuals, similar to the parents. In evolutionary multi-agent system, each agent is equipped with the ability of reproduction. It consists in creating of the offspring agent and passing to it (inheritance) of certain features of the parent, or certain combination of the features of parents in the case of sexual reproduction (recombination) with possibility of small random changes (mutation). Of course besides the inherited features, the agent may possess certain information that are not inherited and that may be changed in the course of its life (e.g. the knowledge about the surrounding environment).

Depending on the assumed evolution model, the features inherited by the agent may be called:

- *genetic material* (the evolution model is similar to the one used in genetic algorithms),
- *individual features* (the evolution model is similar to the one used in evolution strategies),
- *species features* (the evolution model is similar to the one used in evolutionary programming).

In the first case, the basic parameters depicting the agent behavior become its genotype—they are inherited from its parents in the moment of creation and remain unchanged during all of its life. In this case, the complete information—inherited and acquired—describes to the full extent the behavior of the agent in the environment—it phenotype. In the second and the third case the inherited features describe the agent behavior in full.

At the same time the reproduction is directly connected with the position of the agent in the environment:

- the newly created agent is placed in the nearest vicinity of the parent or parents,
- in the case of recombination the parents must be located in their neighborhood,
- the evaluation of the individual may be realized only in comparison to its neighboring agents.

The possibilities of migration, meant as searching for appropriate location for agent in the environment, will have in EMAS a significant influence on the course of evolutionary processes.

2.3.2 Selection

The biggest problem in the evolution process assumed in EMAS is the way of realization of the second crucial element of the evolutionary processes—the natural selection rule. Because of autonomy of the individuals (agents), one cannot simply realize the mechanisms known in the classic evolutionary algorithms. At the same time, a very important aspect of the evolutionary agent system that must be considered during the realization of the selection mechanisms is the dynamics of the population. The selection process cannot lead to significant reduction of the agents, at the same time should not lead to their excessive growth in number. The first problem leads to stagnation of the evolution process, or even to the extinction of population, the second one to decreasing of the efficiency of the system.

In the approach discussed here, following the mechanisms observed in nature, the selection process is realized based on the described in the previous section mechanisms based on non-renewable resources. In the simplest case, one kind of such resource is used, called *life energy*. Each of agents is assigned certain levels of the life energy, that is necessary for performing certain actions, in particular connected to the evolution process, such as reproduction. The agent may undertake the decision to realize the reproduction action, when the level of its energy exceeds the *reproduction threshold*. Thanks to this only the best individuals, on the basis on their energetic status, have the highest probability of reproduction. At the same time, agents having low energy should be successively eliminated from the population. In order to realize this in the basic model of EMAS, a *death threshold* is set. If the energy level of an agent falls below the death threshold, the agent is removed from the population—a typical value of the death threshold is 0, so the agent is removed from the population when it looses the whole life energy. As a consequence, in the system there remain only the agents with the best quality (from the point of view of the problem to be solved), and by the manipulation of the available energy, the size of the population can be controlled.

In the beginning of the evolution process, each agent receives the same energy portion $e_i[t_0] = e_{init}$ (e_i means here the energetic level of the i-th agent, t_0 means the moment of its creation). This value along with the population size are very crucial parameters describing the features of the system. Similarly, a certain energy portion (dependent on the parent or the environment) is passed to the agent created in the process of reproduction. This value should be of course lower than the reproduction threshold, and higher than the death threshold ($e_{repr} > e_i[t_0] > e_{death}$). Later, in the course of agent's life, its energy rises or falls down caused by the actions it realizes:

$$e_i[t+1] = e_i[t] + \Delta e_i^+[t] - \Delta e_i^-[t] \tag{2.1}$$

where $\Delta e_i^+[t]$ is the sum of the energy gained and $\Delta e_i^-[t]$ is the sum of energy lost by the i-th agent in the t-th step.

As a consequence, the agent can gain or loose the possibility of realizing of certain actions, in particular the action of reproduction, or even, when the energy level becomes lower than the death threshold, it will be removed from the system. Assuming that each change of the energetic level of the agent is accompanied by a reversed change of the energy of a certain agent or the environment, one can achieve the effect of maintaining the constant value of the total energy in the whole system, forming so called *energy preserving rule*:

$$\sum_{i=1}^{N} e_i[t] = const \tag{2.2}$$

where N stands for the number of the agents in the population at the time t.

This feature makes possible (using the parameters of the dynamics of the evolutionary processes—death and reproduction thresholds) assessing the limit of the agent population.

The most important mechanism of the inter-agent energy flow is the meeting strategy. Each agent in each time step undertakes (usually randomly) a decision of choosing another agent in order to compare the quality of their solutions (from the optimization criterion point of view), and if such comparison is possible, the better agent receives a predefined portion of energy from the worse agent (such setting resembles somehow the tournament selection well-known in the classic evolutionary algorithms). Thus the agents representing the best solutions in the population in the subsequent steps will increase their energy level, overtaking the other agents' energy, that in the consequence would lead to reaching certain energy level and to reproduction. At the same time the agents representing the lowest quality solutions in the population will successively loose their energy, leading to reaching the predefined death threshold, and to their elimination from the system. Thus the meeting strategy along with the death and reproduction actions realize the goal of the selection, i.e. preference of the best solutions from the point of view of the optimization criterion.

The energy must be also localized in the environment, that may participate in the process of the energy exchange among the agents. In the simplest case the environment may acquire the energy from the removed agents and pass them to the newly created agents. It can also participate in the evaluation mechanisms and assign the agents the energy in such way, that the better agents get more than the worse ones.

2.3.3 Computational Features

In the case of applying evolutionary multi-agent system as a technique for solving search and optimization system, as a method closer to the natural evolution model, one can expect the following advantages of the mechanism applied:

- local selection allowing for creation of geographical niches, meaning more intensive exploration of different search space parts at the same time,
- evolving of phenotype during the life of the individual (agent) based on its interaction with the environment,
- self-adaptation of the population size to the solved problem, applying appropriate selection mechanisms, allowing for efficient management of the population dynamics.

Introduction of the environment space along with the migration mechanisms makes easier the implementation of the system in the distributed environment, that should ease its efficiency.

Local Selection

As it was said before, one of possibility brought by the EMAS model is distribution of the evolution process in the environment of the agent system, and as a consequence, realization of actions by the agents in the context of other agents and resources in certain location only. Thus the realization of the selection mechanisms in EMAS is a consequence of the assumption of the lack of global knowledge and synchronization, that must be applied in each agent system (cf. Sect. 2.2.1). It makes impossible the evaluation of all the agents at the same time, followed by the creation of certain niches, where the subpopulation may concentrate in different parts of the search space.

This mechanism reminds of co-evolutionary techniques or parallel evolutionary algorithms (cf. Sect. 1.2.4), one can expect here similar advantages as offered by them. First of all the diversity of the population should increase, followed by the so-called convergence reliability [12], that becomes very important in the practical applications of EMAS as a technique of solving the problems with unknown characteristics. The mechanisms supporting of the creation of niches make this approach particularly useful for solving multi-modal and multi-criteria problems (cf. [244]).

Evolving of Phenotype

In the case of applying the genetic evolution model (similar to genetic algorithms), one can see the subsequent characteristic feature of EMAS. The agent becomes here the object of the evolution process and the entity functioning in its environment, that makes possible its adaptation (learning) in the course of its life to the encountered conditions, making possible development of not-inherited features, being the part of phenotype. As an example of using this feature, one can consider applying of local optimization techniques, enhancing the solution owned by the agent in the course of its life [49]. One can also consider this as a feature particularly useful for the problems, based on searching for the optimal parameters of the algorithms that may be realized at the same time as evolution process (e.g. search for optimal parameters of neural network architecture, where the genotype consists of the network parameters and the phenotype—the network trained in the course of agent's life).

2.3.4 Maintaining Population Diversity

The greatest advantage of the above-described, basic variant of EMAS is the simplicity of its realization, accompanied by its rich characteristics in the case of solving search and optimization problems. Similarly as in the case of classic evolutionary computing, one can identify a broad spectrum of problems, for which these features are insufficient. One can point out the applications requiring higher adaptive abilities, such as in the case of non-stationary environment, which characteristics changes in time, or in the case of multi-criteria optimization. It turns out that the EMAS model can be treated as a convenient base for natural implementation of many mechanisms observed in nature, which—sometimes with great difficulty—are used in the classic evolutionary computing and other population-based techniques of computational intelligence [186].

An example of a simple realization of a classic concept from the group of niching and speciation techniques, is the implementation of the crowding, using the meeting strategies and the energy transfer. It makes a counterweight to the observed in all evolutionary techniques preference to group the solutions in the local extrema basins of attraction, making the main cause for loosing of diversity of the solutions in the population. This is realized by decreasing of the energy of the individuals representing very similar solutions, also these potentially populating the same ecological niche (i.e. the basin of attraction of the same local extremum).

In the considered variant of the meeting strategy, besides the comparison of the quality, also the distance between the individuals is computed, according to the assumed metric, e.g. for the continuous problem realized as follows:

$$d(x^A, x^B) = \sum_{i=1}^{N} |x_i^A - x_i^B| \qquad (2.3)$$

where $x^A = [x_1^A, \ldots, x_N^A]$ i $x^B = [x_1^B, \ldots, x_N^B]$ are solutions represented by the meeting agents. The work of the mechanism of the *energetic crowding* depends on the value of the parameter ξ called *crowding factor*. If the similarity (distance) between the solutions exceeds certain value of this parameter, the agent initiating the meeting overtakes the energy from the encountered agent in the value of Δe depending on the owned energy and on the level of similarity between the solutions:

$$\Delta e = \begin{cases} 0 & \text{if } d \geq \xi \\ e_B \cdot \left(1 - \frac{d^2}{\xi^2}\right) & \text{in other cases} \end{cases} \qquad (2.4)$$

where e_B is the energy level owned by the encountered agent and d is the computed similarity (distance) between the solutions.

The energy flow connected with the meeting mechanism makes possible, that too similar (considering the solutions owned) agents will loose their life energy in their subsequent life steps, leading to restricting their reproduction ability and even leading to their removal from the population. Indirectly it may result in increasing the probability of reproduction of the other agents, assuming they will represent good solutions and they will obtain energy in the basic mechanism of meeting strategy (i.e. by comparing their quality). In the single-criterion optimization, this mechanism can be too weak for efficiently maintain the local population in different basins of attraction of the local extrema, that differ to a great extent with the fitness value, but it should work well in the multi-criteria problems (see Sect. 2.4.4).

2.4 Other Variants of EMAS

In this section other, more sophisticated versions of evolutionary multi-agent systems are presented, namely immunological, co-evolutionary and multi-criteria and memetic ones.

2.4.1 Memetic EMAS

EMAS may be held up as an example of a cultural algorithms, where evolution is performed at the level of relations among agents, and cultural knowledge is acquired from the energy-related information. This knowledge makes it possible to state which agent is better and which is worse, justifying the decision about reproduction. Therefore, the energy-related knowledge serves as situational knowledge (see Sect. 1.3.2). Memetic variants of EMAS may be easily introduced by modifying evaluation or variation operators (by adding an appropriate local-search method).

The idea of memetic EMAS consist in putting together a local search technique and the evaluation or variation operators utilized in EMAS (see Fig. 2.7). Therefore, implementation of Baldwinian and Lamarckian memetics in EMAS may be easily carried out in the following way:

- Baldwinian memetics: this implementation is done in much the same way as in classical evolutionary computing where in the course of evaluation of a a certain individual, the actual returned fitness is computed for one of its potential descendants (after running the local search procedure in the genotype domain, starting from the evaluated individual). Usually this is an iterative process that involves

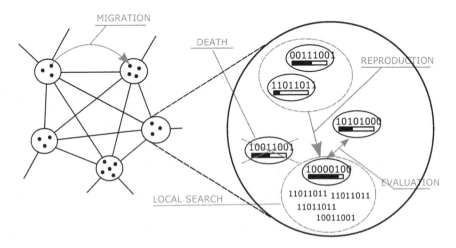

Fig. 2.7 Concept of memetic evolutionary multi-agent system (memEMAS)

many mutations, or using any other dedicated local search technique. The result returned is the fitness of the best encountered genotype, while the genotype of the evaluated individual remains unchanged.

- Lamarckian memetics: agent may improve its genotype (by running a local search procedure in the genotype domain, starting from the evaluated individual, similarly as in the case of Baldwinian memetics). Usually this is an iterative process, involving many mutations, or using any other dedicated local search technique. The result returned is the best encountered genotype. In the case of Lamarckian memetics, the genotype of the evaluated individual is changed. This procedure may be performed either during the reproduction or at the arbitrarily chosen moments of agent's life, depending solely on agent's own decision.

2.4.2 Immunological EMAS

Solving difficult search problems with population-based approaches, especially those with costly evaluation of their candidate solutions (e.g. inverse problems [9]), requires looking for techniques that may make it possible to increase the search efficiency. One method can be reducing the number of fitness function calls. It may be done in several ways, e.g. by applying tabu search [7] or, strictly in a technical layer, by caching fitness values generated for identical (or similar) genotypes. In this section, an immunological selection mechanism designed for EMAS is discussed.

The main idea of applying immunological inspirations to speed up the process of selection in EMAS is based on the assumption that "bad" phenotypes come from "bad" genotypes. Thus, a new group of agents (acting as lymphocyte T-cells) may be introduced (see works of Byrski and Kisiel-Dorohinicki [44, 45] and Ph.D. thesis of Byrski [39]). They are responsible for recognizing and removing agents with genotypes similar to the genotype pattern possessed by these lymphocytes. Another approach may introduce specific penalty applied by T-cells for recognized agents (certain amount of the agent's energy disappears) instead of removing them from the system. The general structure of iEMAS (immunological EMAS) is presented in Fig. 2.8.

Of course, there must exist some predefined affinity (lymphocyte-agent matching) function which may be based, e.g. on the percentage difference between corresponding genes. Lymphocytes are created in the system after the action of death. The late agent genotype is transformed into lymphocyte patterns by means of mutation operator, and the new lymphocyte (or a group of lymphocytes) is introduced into the system.

Antibodies in the real immune system of the vertebrates are subjected to the *negative selection* process in thymus, where they test other cells belonging to the organism—if during this time they recognize the "self" cell, they are considered as infeasible and removed. This mechanism has to counteract too quick realization of the removing of the antigens, that may turn out to be "self" cells indeed. Within a specified period of time, the affinity of immature lymphocytes' patterns towards

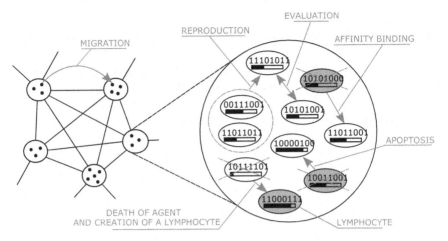

Fig. 2.8 Structure and the principle of work of immunological variant of EMAS (iEMAS)

"good" agents (possessing relatively high amount of energy) is tested. If it is high (lymphocytes recognize "good" agents as non-self), they are removed from the system. If the affinity is low, it is assumed that they will be able to recognize non-self individuals ("bad" agents) leaving agents with high energy intact.

The antibodies are subject of the *apoptosis* process, that is implemented in order to retain the capability of forgetting and adaptation by the immune system. A similar process in realized in the described agent system. The lifespan of lymphocytes is controlled by specific, renewable resource (strength), used as a counter by the lymphocyte agent, and with its depletion the immune cell is removed from the system.

Summing up, the cycle of the lymphocyte agent (T-cell) is realized as follows:

1. During removal of the computing agent, in the system there are created one or more of the T-cells. Each of these cells obtains the pattern of the solution of the removed agent (it can be mutated though). Thanks to this the T-cells become a vessel for potentially "bad" solutions.
2. Immature T-cells test the agents for some time (realizing negative selection), and if they recognize an agent with relatively high level of the energy, they are removed from the system. After finishing this process they work in an unconstrained manner.
3. T-cell owns a pattern of a potentially "bad" solution of the problem (its structure is identical as the structure of the genotype) and uses certain probability measure seeking similar solutions in the system, in order to remove them or to penalize the agents that own these solutions (in the simplest case this "similarity" or "affinity" function may be realized as Euclidean distance in the domain of continuous optimization, or Hamming distance in discrete optimization).
4. T-cell after finishing its negative selection process may work in the system during a certain period of time. After finishing this, it is removed from the system, however

this time may be extended by rewarding the T-cell for proper recognitions of the agents.

iEMAS is an example of a cultural algorithm, which is an extension of EMAS. However, based on the agent energy-related information the population of lymphocytes is modified (they are considered mature or not). Therefore, the agent energy-related knowledge serves as situational knowledge (see Sect. 1.3.2). Lymphocyte energy serves as temporal knowledge, so the lymphocyte may be removed after a certain period of time.

Removing from the population the solutions infeasible or "bad" (from the point of view of the current goal) can be perceived as similar to the tabu search algorithms [8]. However it should be noted, that in the proposed approach there is no a globally accessible set of "tabu" solutions, instead the introduced T-cells act locally, searching for infeasible solutions in its neighborhood. This can be perceived as a somehow "softened" tabu search.

The early removing of "bad" solutions and a decrease in the number of the individuals in the computing populations (see the computing results later in this section) makes iEMAS a weapon of choice to deal with problems, where a complex fitness function is used. Indeed, an interesting optimization task was approached with iEMAS, namely the evolution of neural network architecture [44, 47] and benchmark function optimization [45].

2.4.3 Co-evolutionary Multi-agent Systems

The concept of introducing into EMAS man co-evolving species (or sexes) [90, 91] is aimed at maintaining a significant diversity of the population by developing subpopulation localized in the basins of attraction of the local extrema, as it is realized in the niching and speciation techniques. This approach can also be used for solving multi-modal optimization problems.

The starting point for the further considerations on *co-evolutionary* multi-agent systems (CoEMAS) is the observation, that similarly to the selection mechanism, it is quite difficult to apply classic co-evolutionary techniques in EMAS, because of the implicit characteristics of the multi-agent systems (cf. Sect. 2.2.1). At the same time one can use the existing concepts connected with existence of the system space and resources in order to introduce allopatric speciation and competition for the limited resources, and decentralized characteristics of the interaction among the agents may easily depict the existing influences between the species (or sexes).

The simplest solution leading to of allopatric speciation is introduction of high cost (meant as loosing the life energy) of migration among the islands [91]. The speciation process is based in this case on the geographical isolation of sub-populations, that may have chance to group the solutions from the neighborhoods of the basins of attraction of different local extrema, though in this case there exists no mechanism that could counteract localization of the same extrema by different populations.

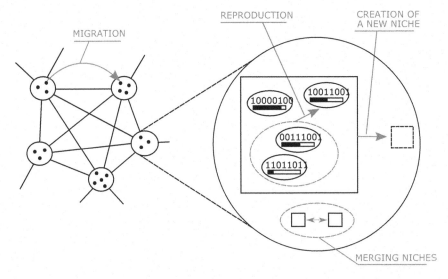

Fig. 2.9 Co-evolutionary multi-agent system (CoEMAS) with co-evolution of species

In the CoEMAS with *co-evolution of species* there exist two types of agents and in fact, two organization levels of the system: agents representing solutions live "inside" the *niche* agents, so they divide the population into two isolated (from the reproduction point of view) populations (see Fig. 2.9). These niches interact among themselves (a competition for limited resources may occur, migration among the islands and merging of niches) and the solutions (by the migration of solutions among the niches), so their solutions are localized in the basins of attraction of different local extrema.

CoEMAS with *a sexual selection* allows for existing of agents of many species and inside the species—many sexes on the same organizational level of the system. The competition for the limited resources along with the phenomenon of conflict and co-evolution of sexes (higher reproduction costs in the case of females that prefer the partners localized in the basin of attraction of the same local extremum) lead to sympatric speciation—creation of groups of agents (species), carrying solutions from the basins of attraction of different local extrema [93].

Application of *mutualism* mechanism (cooperation) leads to introducing into the system several cooperating species of individuals, which cooperation may consist in e.g. solving sub-problems of the same problem. Thus the complete group of all individual belonging to all species may produce the complete solution of the considered problem [94, 98].

In the co-evolutionary multi-agent system with *predator-prey* mechanism, there exist two species of agents: the prey have encoded in the genotypes the solution of the problem, while the predator eliminate from the system the prey having relatively low level of energy [100].

In the all above-referenced variants, the mechanisms applied support stable maintaining of the higher diversity of the solutions, giving similar effects as the ones observed in classic co-evolutionary techniques.

2.4.4 EMAS in Multi-criteria Optimization

A very interesting application of the concept of evolutionary multi-agent system is the task of multi-criteria optimization in the Pareto sense. The goal of the system is in this case finding a set of solutions that may be treated as a certain approximation of the Pareto front for the considered poly-optimization problem. In the simplest case, evolving population of agent may represent the set of feasible solutions of the problem described by the embedded into the system criteria functions. The most crucial element of the evolution process in this case is the meeting strategy, realized in the case of poly-optimization according to the dominance relation. In this way the agents representing the non-dominated solutions gain the energy and may undertake the reproduction process with higher probability. At the same time the dominated agents loose their energy, and in the course of time are eliminated from the system. Because of that, in the subsequent evolution stages, the set of the non-dominated agents should represent subsequent approximations of the Pareto front of the considered problem [83, 176].

Unfortunately, EMAS, similarly to other evolutionary algorithms, requires additional mechanism for maintaining diversity of the population on such level, that possibly uniform sampling of the whole Pareto front could be realized. One of ways that can be applied is taking advantage of the crowding technique (sf. Sect. 2.3.4). This mechanism is effective especially effective in the case of solving the problems with non-uniform Pareto front, because of more intensive exploration of the search space [170, 178, 268].

At the same time the described process may lead to the stagnation phenomenon (lack of directed evolution), when after passing a certain time, the population is composed of nearly agents representing nearly identical (mutually non-dominated) solutions, causing loss of energy flow during meetings. In this case one can try to apply elitism mechanisms that can be realized in EMAS in the variant described as archive.

Elitist Variant

It is noteworthy, that in the context of general understanding of elitism as an algorithmical feature, consisting in preserving the already found solutions, the above-described basic EMAS variant may be perceived as elitist. This is because the selection based on resources allows any agent to remain in the system, until it encounters sufficient number of better agents, that during the energy transfers its energy level will fall down below the death threshold. In order to differentiate the presented elitist variants, this one is called *semi-elitist* one.

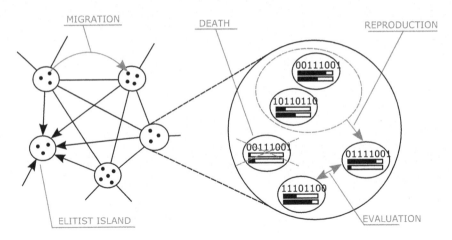

Fig. 2.10 Concept of elitist evolutionary multi-agent system (elEMAS)

The concept of *elitist* evolutionary multi-agent system (elEMAS) consists in introducing so-called *elitits* island [262], into the computing system structure, that becomes a certain archive (see Fig. 2.10). On the elitist island the agents do not confer to "regular" evolution rules, realizing only a special, predefined meeting strategy. If during the meeting it turns out, that the solution represented by one of the agents dominates the solution of the other agent, the dominated agent performs the death action at once. Thus one can assume that on the elitist island, only non-dominated agents may be found, i.e. representing the approximation of the Pareto front.

A crucial assumption which must me made here is that only the agents representing high-quality solutions are transferred to the elitist island. In order to make such assessment possible, in elEMAS an additional resource was introduced, so-called *prestige*, that is gained by the agent if it dominates another agent during the meeting realized on one of regular islands. If the agent passes so-called elitist threshold (measured by the prestige resource) it means that compared to other agents, its solution is valuable and it can be transferred to the elitist island. The final decision on this transfer is undertaken by the agent based on the knowledge about other similar solutions—the migration is possible only if the agent has identified sufficient number of similar solutions (this may be compared to crowding mechanism. A side-effect of this decision is elimination of such agent from the evolution process, thus increasing the chances of other agents that should enhance the diversity of the population [264–266].

Co-evolutionary Techniques

The applications of the EMAS to multiobjective optimization encompass also the previously mentioned co-evolutionary variants—using parasitism [96], mutualism [101], predator-prey [97] and sexual selection [95, 99].

2.5 Summary

In this chapter a review of the basic issues connected with structure and behavior of the described class of computing systems hybridizing the evolutionary and agent-oriented paradigms were presented. The authors aimed at showing the simplicity and features of the basic version of evolutionary multi-agent system (EMAS), and at the same time, its plethora of proposing potential extensions, following different nature-based inspirations. Because of that, besides basic EMAS, in this section different its extensions were presented (immunological, co-evolutionary, elitist and memetic).

The developed new metaheuristic algorithm has been present in publication from 1996. However after recent influential paper by Sörensen [270]. We would like to claim that EMAS directly hybridizes agent-based and evolutionary paradigms, utilizes in full the already present terminology and tackles many different areas and problems. Therefore as far as we know the makers of EMAS (to whom we also belong) did their best not to fall into category of so-called "novel" metaheuristics identified and fought against by Sörensen.

Decentralization of the evolution process can yield considerable advantages in application to solving certain problem classes and using particular mechanisms [43]. However besides the presented computational features of the techniques, the authors are far from claiming that they are the best, universal optimization "solvers". One has to still remember about the well-known *no free lunch theorem* [297], stating, that the best solutions are obtained by using a method dedicated for solving certain classes of problems.

The presented outline of different agent-based evolutionary computing concepts does not cover many aspects of the total characteristics of the presented problems. On this stage the main goal was to formulate general principles of their structure and behavior and sketching out the domain of their applications. At the same time many additional issues may arise, connected with analysis and design of such complex systems (such as implementation features or formal analysis). These will be tackled in the next sections.

Chapter 3
Formal Aspects of Agent-Based Metaheuristics

As stated in the previous chapter, agent-based metaheuristics, such as EMAS and iEMAS have already proved to be good techniques for solving difficult search problems (see, e.g. [44–47, 92, 263] to point out a few). However, in order to make sure that such a complex tool may be useful, both formal anadeflysis and experimental verification should be conducted.

The model presented here was not only constructed for the sole purpose of EMAS, but also to constitute a framework that may be further used to describe other computing systems (therefore, EMAS model was extended in order to support iEMAS). Research may be continued, and other similar computing systems may be defined using this style of modeling.

Using these models for the sole purpose of definition of computing systems will not make it possible utilise its descriptive power to the full extent. In the next chapter, a detailed analysis of EMAS stochastic features based on the model presented in this chapter is given. The Markov-chain modeling EMAS has turned out to be ergodic, which gives for this class of systems asymptotic guarantee of success (cf. [241, 253]). For iEMAS, appropriate ergodic conjecture is given and the proof is outlined.

The formal model of EMAS together with the ergodicity proof may be seen as unique phenomenon in the field of computing, as there is a lack of a comprehensive stochastic model of the wide class of population-based, general evolutionary or memetic algorithms.

The research on modeling and formal analysis of EMAS and iEMAS was conducted in close cooperation with Robert Schaefer and Maciej Smołka.

3.1 Formal Definition of EMAS

The structure and behavior of EMAS is described in this section, according to a model presented in [255] that was refined and extended in [51].

Computing agents in EMAS may be seen as autonomous individuals. Each agent is capable of observing its environment by gathering important information, making

© Springer International Publishing AG 2017
A. Byrski and M. Kisiel-Dorohinicki, *Evolutionary Multi-Agent Systems*,
Studies in Computational Intelligence 680, DOI 10.1007/978-3-319-51388-1_3

decisions which affect its activity, and performing actions leading to introducing changes in the system state see, e.g. [154, 155].

Each agent is assigned a genotype that belongs to the finite universum U, $\#U = r < +\infty$ which can be the a set of binary strings, real numbers or other codes providing basis for solving particular global optimization problem. Agents are assigned to locations (islands) and can migrate between them.

Genetic operations performed on agents' genotypes, such as crossover and mutation, are practically similar to those used in classical evolutionary algorithms and lead to creating a new agent (see Sect. 3.1.2). The EMAS agent can also create its offspring, using the predefined action of cloning with mutation.

As described in the previous section, the selection mechanism is based on the existence of a non-renewable resource called *life energy*, which is gained and lost when agents perform actions (see [169]).

EMAS has the following characteristic features:

- *Quasi-signature of an agent*: it is composed of its (invariant) genotype *gen* and the numerical identifier of the copy n that is changed during the migration.
- *Fitness function*: the function $\psi : U \to [0, M]$ related in some way to the objective Φ where again $\mathbb{R}_+ \ni M < +\infty$. In the simplest case $\psi(gen) = \Phi(\eta(gen))$, where $\eta : U \to \mathcal{D}$ is the decoding function.
- *Variable location of agents*: active EMAS agents are present in locations described by a set of immutable integer labels $Loc = \{1, \ldots, s\}$. The locations are linked together by channels along which agents may migrate from one location to another. The symmetric relation $Top \subset Loc^2$ determines the topology of locations. It is assumed that the connection graph $\langle Loc, Top \rangle$ is coherent and does not change during the system evolution.
- *Dynamic collection of agents*: agents belong to the predefined finite set Ag, which can at every moment be one-to-one mapped into the set $U \times P$, where $P = \{1, \ldots, p\}$ and p is assumed to be the maximal number of agents containing the same genotype. In other words, each agent $ag_{gen,n} \in Ag$ contains one potential solution to a given problem encoded as $gen \in U$. More than one agent may be present in the system containing this solution and the index $n \in P$ is used to distinguish them. Furthermore, it is assumed that each location has its own separate subset of admissible genotype copy numbers P_i, i.e. $P = \bigcup_{i \in Loc} P_i$, $P_i \cap P_j = \emptyset$ for $i \neq j$ and $n \in P_i$ as long as the agent with the temporary copy number n resides at i.
- *Variable energy of agents*: each agent may possess only one of the following quantized energy values: $0, \Delta e, 2 \cdot \Delta e, 3 \cdot \Delta e, \ldots, m \cdot \Delta e$.

Although a pair (gen, n) is not a true identifier because of the variability of its second component n, it properly distinguishes different agents at any time-moment. Reference to a certain agent will be given as: *agent* $ag_{gen,n}$ remembering that this notation is time-dependent. Note that in the context of finding the objective function minimizers, the identity of agents (i.e. solution holders) is less important. Thus the crucial agent attributes are the genotype (and the fitness as its derivative) and the life energy (see the sequel), whereas the copy number plays only an auxiliary role.

On the other hand, if this approach was to be aligned one with the Belief-Desire-Intention (BDI) model [238], a true agent identifier could be constructed by means of a quasi-signature as the composition of the agent's genotype with the sequence of the agent's copy numbers at subsequent moments, putting 0 at the moments when the agent is active. However, it is worth noting that such an identifier would not be very useful unless it was stored in a globally synchronized repository, and the need of the global synchronization would, in turn, prevent the concurrent performance of some crucial EMAS actions.

3.1.1 EMAS State

The following three-dimensional incidence and energy matrices $x \in X$ with s layers (corresponding to all locations) $x(i) = \{x(i, gen, n), \; gen \in U, \; n \in P\}$, $i \in Loc$ will provide a basis for the description of the system state. The layer $x(i)$ will contain energies of the agents in the i-th location. In other words, the condition $x(i, gen, k) > 0$ means that the k-th clone of the agent carrying the gene $gen \in U$ is active, its energy equals $x(i, gen, k)$ and it may be found in i-th location.

Now, the following coherency conditions are introduced:

- $(\cdot, j, k)$-th column contains at most one value greater than zero, which means that the agent with the k-th copy of the j-th genotype may be found in only one location at a time, whereas other agents containing copies of the j-th genotype may be present in other locations;
- entries of incidence and energy matrices are non-negative $x(i, j, k) \geqslant 0$ for $1 \leqslant i \leqslant s, 1 \leqslant j \leqslant r, 1 \leqslant k \leqslant p$ and $\sum_{i=1}^{s} \sum_{j=1}^{r} \sum_{k=1}^{p} x(i, j, k) = 1$, which means that the total energy of the whole system is constant and equal to 1;
- $x(i, gen, n)$ can be positive only for n acceptable in the location i, i.e. $n \in P_i$;
- each layer $x(i)$ has at most q_i values greater than zero, denoting the maximal capacity of the i-th location and, moreover, the quantum of energy Δe is less than or equal to the total energy divided by the maximal number of individuals that may be present in the system,

$$\Delta e \leqslant \frac{1}{\sum_{i=1}^{s} q_i}, \tag{3.1}$$

which makes it possible to achieve the maximal population of agents in the system;
- reasonable values of p should be greater than or equal to 1 and less than or equal to $\sum_{i=1}^{s} q_i$; it is assumed that $p = \sum_{i=1}^{s} q_i$, which assures that each configuration of agents in locations is available, in respect of the total number of active agents $\sum_{i=1}^{s} q_i$; increasing p over this value does not enhance the descriptive power of the presented model;
- the maximal number of copies for each location #P_i should not be less than q_i in order to make possible for the system to reach a state in which particular location

is filled with clones of one agent; on the other hand due to the previous assumption $\#P_i$ cannot be greater than q_i; therefore it is finally assumed that $\#P_i = q_i$.

Gathering all these conditions, the considered set of three-dimensional incidence and energy matrices constituting the EMAS space of states may be described:

$$X = \left\{ x \in \{0, \Delta e, 2 \cdot \Delta e, 3 \cdot \Delta e, \ldots, m \cdot \Delta e\}^{s \cdot r \cdot p}, \right.$$

$$\text{subject to: } \Delta e \cdot m = 1, \sum_{i=1}^{s} \sum_{j=1}^{r} \sum_{k=1}^{p} x(i, j, k) = 1,$$

$$x(i, j, k) = 0 \text{ for } 1 \leqslant i \leqslant s, 1 \leqslant j \leqslant r, k \notin P_i,$$

$$\sum_{j=1}^{r} \sum_{k=1}^{p} [x(i, j, k) > 0] \leqslant q_i \text{ for } 1 \leqslant i \leqslant s,$$

$$\left. \sum_{i=1}^{s} [x(i, j, k) > 0] \leqslant 1 \text{ for } 1 \leqslant j \leqslant r, 1 \leqslant k \leqslant p \right\} \qquad (3.2)$$

where $[\cdot]$ is the indicator function, i.e. $[\texttt{true}] = 1$ and $[\texttt{false}] = 0$.

Structure and Behavior of EMAS

EMAS may be modeled as the following tuple:

$$\langle U, Loc, Top, Ag, \{agsel_i\}_{i \in Loc}, locsel, \{LA_i\}_{i \in Loc}, MA, \omega, Act \rangle, \qquad (3.3)$$

where:

MA   (Master Agent) is used to synchronize the work of locations; it allows to perform actions in particular locations; this agent is also used to introduce the necessary synchronization into the system;

$locsel : X \to \mathcal{M}(Loc)$ is a function used by *MA* to determine which location should be permitted to perform the next action;

LA_i (Local Agent) is assigned to each location; it is used to synchronise the work of Computing Agents (*CA*) present in the location; LA_i chooses the *CA* and allows it to evaluate a decision and perform the action, at the same time requesting permission from *MA* to perform this action;

$agsel_i : X \to \mathcal{M}(U \times P)$ is a family of functions used by LA_i to select the *CA* that may perform the action, such that each location $i \in Loc$ has its own function $agsel_i$; probability $agsel_i(x)(gen, n)$ vanishes when the agent $ag_{gen,n}$ is inactive in the state $x \in X$ or is present in location different from i-th one ($n \notin P_i$);

$\omega : X \times U \to \mathcal{M}(Act)$ is a function used by agents to select actions from the set *Act*; both symbols will be explained later.

Act   is a predefined, finite set of actions.

Hereafter $\mathcal{M}(\cdot)$ stands for the space of probabilistic measures. Moreover, the following convention will be used to describe the discrete probability distributions

$a : X \times Par \rightarrow \mathcal{M}(A)$ on the set A, depending on the state $x \in X$ and an optional parameter $p \in Par$.

- $a(x, p)(d)$ is the probability of $d \in A$ (lower-case letter), assuming the current state x and the parameter p. It simplifies a more rigorous notation $a(x, p)(\{d\})$.
- $a(x, p)(W)$, $W \subset A$ is the probability of set W (upper-case letter).

The names a, A, Par are of course generic. The variables d and p may also be defined as tuples. In the case of probability distribution on the finite set A, this notation is unambiguous.

The population of agents is initialized by means of introductory sampling. This may be regarded as a one-time sampling from X, according to a predefined probability distribution (possibly the uniform one) from $\mathcal{M}(X)$. Each agent starts its work immediately after being activated. At every observable moment only one agent present in each location gains the possibility of changing the state of the system by executing its action.

The function $agsel_i$ is used by LA_i to determine which agent present in the i-th location will be the next one to interact with the system. After being chosen, the agent $ag_{gen,n}$ chooses one of the possible actions according to the probability distribution $\omega(x, gen)$. Note that there is a relationship of this probability distribution and the concept of fine-grain schedulers introduced into a syntactic model of memetic algorithms in [183]. It must be noted that the selection of action by all agents, which carry the same genotype gen in the same state x, is performed according to the same probability distribution $\omega(x, gen)$ and does not depend on the genotype copy number n. In the simplest case, ω returns the uniform probability distribution over Act for all $(x, gen) \in X \times U$.

Next, the agent applies to LA_i for the permission to perform this action. When the necessary permission is granted, the agent $ag_{gen,n}$ performs the action after checking that a condition defined by formula (3.5) has been fulfilled. If during the action the agent's energy is brought to 0, this agent suspends its work in the system (it becomes inactive).

MA manages the activities of LA_i and allows them to grant their agents permission to carry out requested tasks. The detailed managing algorithm based on the rendezvous mechanism [145] is described in Sect. 3.1.3.

A subset of states in which the agent $ag_{gen,n}$ is active is denoted as:

$$X_{gen,n} = \{x \in X \mid \exists\, l \in Loc : x(l, gen, n) > 0\},\ (gen, n) \in U \times P \qquad (3.4)$$

Each action $\alpha \in Act$ will be represented as a pair of function families $(\{\delta_\alpha^{gen,n}\}_{(gen,n)\in U\times P}, \{\vartheta_\alpha^{gen,n}\}_{(gen,n)\in U\times P})$. The functions

$$\delta_\alpha^{gen,n} : X \rightarrow \mathcal{M}(\{0, 1\}) \qquad (3.5)$$

make it possible to take a decision, to perform the action. The action α is performed with probability $\delta_\alpha^{gen,n}(x)(1)$ by the agent $ag_{gen,n}$ at state $x \in X$ and rejected with probability $\delta_\alpha^{gen,n}(x)(0)$. Because the action may be invoked only by the active agent, the function $\delta_\alpha^{gen,n}$ always has to return a negative decision for all $x \in X \setminus X_{gen,n}$ and only the restriction $\delta_\alpha^{gen,n}|X_{gen,n}$ constitutes the crucial part of this function, so

$$\delta_\alpha^{gen,n}(x) = \begin{cases} \delta_\alpha^{gen,n}|X_{gen,n} & x \in X_{gen,n} \\ (1,0) & x \in X \setminus X_{gen,n}. \end{cases} \quad (3.6)$$

Next, the formula

$$\vartheta_\alpha^{gen,n} : X \to \mathcal{M}(X) \quad (3.7)$$

defines non-deterministic state transition functions, therefore $\vartheta_\alpha^{gen,n}$ is caused by executing action α by the agent $ag_{gen,n}$. The value of $\vartheta_\alpha^{gen,n}(x)(x')$ denotes the probability of passing from the state x to x' resulting from the execution of the action α by the agent $ag_{gen,n}$. The function is only invoked if the agent is active, therefore it is enough to define a restriction $\vartheta_\alpha^{gen,n}|X_{gen,n}$ for each action α, and let it take an arbitrary value on $X \setminus X_{gen,n}$.

If any action is rejected, the trivial state transition

$$\vartheta_{null} : X \to \mathcal{M}(X) \quad (3.8)$$

such that for all $x \in X$

$$\vartheta_{null}(x)(x') = \begin{cases} 1 \text{ if } x = x' \\ 0 \text{ otherwise} \end{cases} \quad (3.9)$$

is performed.

The probability transition function for the action α performed by the agent with the n-th copy of the genotype gen

$$\varrho_\alpha^{gen,n} : X \to \mathcal{M}(X) \quad (3.10)$$

is defined by the formula

$$\varrho_\alpha^{gen,n}(x)(x') = \delta_\alpha^{gen,n}(x)(0) \cdot \vartheta_{null}(x)(x') + \delta_\alpha^{gen,n}(x)(1) \cdot \vartheta_\alpha^{gen,n}(x)(x'), \quad (3.11)$$

where $x \in X$ denotes the current state, and $x' \in X$ is the consecutive state resulting from the conditional execution of α.

Note that it is formally possible to consider a very large (yet finite) set *Act*, comprising all actions up to a certain description length (using a Gödel numbering [132] or any appropriate encoding). This implies that this set may be implicitly defined by such an encoding, allowing much flexibility in the set of actions available (a connection can be drawn with multi-meme algorithms [182]).

Observation 3.1.1 ([51, Observation 2.1]) *Given the agent $ag_{gen,n} \in Ag$ it is enough to define two restrictions $\delta_\alpha^{gen,n}|X_{gen,n}$ and $\vartheta_\alpha^{gen,n}|X_{gen,n}$ in order to establish the probability transition function $\varrho_\alpha^{gen,n}$ associated with the execution of the action α— see Eqs. (3.10) and (3.11).*

The agents' actions may be divided into two distinct types:

- global—changing the state of the system in two or more locations, so only one global action may be performed at a time,
- local—changing the state of the system in one location considering only the state of locally present agents; only one local action for one location may be performed at a time.

Therefore, the Act set is divided in the following way:

$$Act = Act_{gl} \cup Act_{loc} \tag{3.12}$$

Informally speaking, if the location $i \in Loc$ contains the agent performing a certain local action $\alpha \in Act_{loc}$, only the entries of layer $x(i)$ of the incidence and energy matrix are changed (other changes have zero probability). Moreover, these actions do not depend on other layers of x. The action $null$ is obviously "the most local one", because it does not change anything at all.

The above description can be formalized as follows:

Definition 3.1 ([51, Definition 3.1]) The action $\alpha \in Act$ is $local$ ($\alpha \in Act_{loc}$) if, and only if, for any agent which can execute the action (i.e., $\forall (gen, n) \in U \times P$) we have:

1. α does not change anything except for part of the state that describes the location l in which $ag_{gen,n}$ is performing the action α (i.e., $x(l)$ being the l−th layer in matrix x), so

$$\forall x \in X : \varrho_\alpha^{gen,n}(x)(x_{next}) = 0, \tag{3.13}$$

for $x_{next} \in X$ such that $\exists i \neq l : x_{next}(i) \neq x(i)$ and x_{next} denotes one of the states which is supposed to be reached at the step immediately after state x appears;
2. α is independent upon any other layers of x which means that

$$\forall x_1, x_2 \in X, x_1(l) = x_2(l),$$
$$\forall x_{1,next}, x_{2,next} \in X, x_{1,next}(l) = x_{2,next}(l)$$
$$\text{and for each } i \neq l$$
$$x_1(i) = x_{1,next}(i), \ x_2(i) = x_{2,next}(i)$$
$$\varrho_\alpha^{gen,n}(x_1)(x_{1,next}) = \varrho_\alpha^{gen,n}(x_2)(x_{2,next}). \tag{3.14}$$

All other actions are considered global (elements of Act_{gl}).

What makes local actions important is the fact that if they are executed by agents present in different locations, they commute. Detailed formal proof of the local actions commutativity is provided in [51].

Local actions must be mutually exclusive within a single location, and global actions are mutually exclusive in the whole system, so only one global action may be performed at a time in the system, but many local actions (one in each location at most) may be performed at the same time.

3.1.2　EMAS Actions

Consider the following set of actions [51]:

$$Act = \{get, repr, migr, clo, lse\}, \qquad (3.15)$$

where *get* lets a better agent take a part of life energy from a worse agent and may make the agent with low energy inactive, *repr* activates the agent as an offspring agent in the system, *migr* denotes migration of agents between two locations, *clo* activates the agent as a mutated clone agent in the system, whereas *lse* allows the local search methods to incorporate into EMAS. Such sample set of actions cover almost all search activities appearing in GA and MA as selection, mutation, crossover and local optimization.

Denote by l the location of the current active agent containing the n-th copy of the genotype *gen* performing the action (i.e., $x(l, gen, n) > 0$). Notice that if the particular state x is established, the location of each active agent is unambiguously determined. Lower index "next" will be used to denote the state which may appear in the next step assuming some current state, e.g. x_{next} is the subsequent possible value of x.

Action Performing Distributed Selection

The energy transfer action *get* is based on the idea of agent rendezvous. Agents meet one of their neighbors (it chooses randomly one of its neighbors—agents from the same location or "island") and during this meeting a quantum of energy Δe flows in direction described by a certain stochastic function *cmp*. The most probable direction is from the worse evaluated agent to the better one, which may be considered a kind of a tournament (see e.g. [210]).

By Observation 3.1.1 the following may be obtained:

Observation 3.1.2 ([51, Observation 6.1]) *The probability transition function* $\varrho_{get}^{gen,n} : X \to \mathcal{M}(X)$ *associated with action get is determined by:*

$$\delta_{get}^{gen,n}|X_{gen,n}(x)(1) = \begin{cases} 1 \text{ if } NBAG_{l,gen,n} \neq \emptyset \\ 0 \text{ otherwise} \end{cases} \qquad (3.16)$$

$$NBAG_{l,gen,n} =$$
$$\{(j, k) : x(l, j, k) > 0 \text{ and } (j \neq gen \text{ or } k \neq n)\} \qquad (3.17)$$

$$\vartheta_{get}^{gen,n}|X_{gen,n}(x)(x') = \frac{1}{\#NBAG_{l,gen,n}} \sum_{(gen,n)\in NBAG_{l,gen,n}}$$

$$\left(cmp(gen, gen')(0) \cdot [x' = next(x, gen, n, gen', n')] + \right.$$

$$\left. cmp(gen, gen')(1) \cdot [x' = next(x, gen', n', gen, n)] \right) \tag{3.18}$$

$$cmp : U \times U \to \mathcal{M}(\{0, 1\}) \tag{3.19}$$

$$next(x, a, b, a', b') = x_{next} :$$

$$x_{next}(i, j, k) = \begin{cases} x(i, j, k) - \Delta e & \text{if } j = a \text{ and } k = b \\ & \text{and } i = l \\ x(i, j, k) + \Delta e & \text{if } j = a' \text{ and } k = b' \\ & \text{and } i = l \\ x(i, j, k) & \text{otherwise.} \end{cases} \tag{3.20}$$

Explanation The decision of the action *get*, $\delta_{get}^{gen,n}$, defined by Eq. (3.16), depends upon the existence of at least one neighboring agent in the same location and is performed by checking the contents of the $NBAG_{l,gen,n}$ set defined by Eq. (3.17). The arbitrary state of the system when the decision is evaluated by the agent $ag_{gen,n}$ is denoted by $x \in X$.

The transition function uses the function *cmp* to compare the meeting agents. This is a probabilistic function that takes advantage of the fitness function ψ in order to compare the agents. The better fitness the genotype has, the greater probability that it will get the quantum of energy from its neighbor. A lower probability is assigned to the reverse flow.

Technically, if $cmp(gen, gen')$ is sampled as 0 then agent with the genotype *gen* increases its energy and the second agent with genotype *gen'* looses its energy. If $cmp(gen, gen')$ takes the sampled value 1, the energy is passed in the opposite direction.

In the case of the positive evaluation of this decision, the state transition described by Eq. (3.18) is performed. This formula comes from Bayes' theorem, which makes us check all possible agents from $NBAG_{l,gen,n}$. For each agent contained in this set a different state transition is performed as described by the function $next(\cdot, \cdot, \cdot, \cdot, \cdot)$. The direction of the energy transfer is determined using the function *cmp*.

The state transition function is constructed according to Eq. (3.20). The incidence matrix $x_{next} \in X$ is obtained from x by changing two entries related to a pair of agents $ag_{gen,n}, ag_{gen',n'}$ that exchanged energy.

Observation 3.1.3 ([51, Observation 6.2]) *The value of the probability transition function imposed by Eqs. (3.10), (3.11), (3.16)–(3.20) performed by the agent $ag_{gen,n}$ present in the location l, depends only on the elements of the system state contained in its location. The action get may only introduce changes in the state entries associated* *with the location l. In other words, assume that x is a current state, all states that differ from x outside the l-th layer have a null probability in the next step.*

Explanation Observation 3.1.3 stems from Eqs. (3.10), (3.11), (3.16)–(3.20). Part of them that introduce changes in the state entries depends on and refers only to the entries in the current l-th location. All other entries are simply rewritten to the next state.

Actions Inspired by the Genetic Operations

A decision on the reproduction action *repr* is based on the idea of the agent rendezvous (similarly to *get*). An agent with sufficient energy (above a certain predefined threshold e_{repr}) meets one of its neighbors and creates offspring agent based on their solutions. The genotype of the offspring agent is selected according to the predefined family of probability distributions $mix : U \times U \to \mathcal{M}(U)$ associated with the sequence of genetic operations (e.g. crossover followed by mutation, see [290]). In particular, $mix(gen, gen')(gen'')$ denotes the probability that gen'' is born of the parents gen and gen'. A part of the parents' energy ($e_0 = n_0 \cdot \Delta e$, n_0 is even) is passed onto the offspring agent.

By Observation 3.1.1 the following may be obtained:

Observation 3.1.4 ([51, Observation 6.3]) *The probability transition function* $\varrho_{repr}^{gen,n} : X \to \mathcal{M}(X)$ *associated with the action repr is determined by:*

$$
\delta_{repr}^{gen,n}|X_{gen,n}(x)(1) = \begin{cases} 1 \ if \ x(l, gen, n) > e_{repr} \ and \ RPAG_{l,gen,n} \neq \emptyset \\ \quad and \ \sum_{j=1}^{r} \sum_{k \in P_l}[x(l, j, k) > 0] < q_l \\ 0 \ otherwise \end{cases} \tag{3.21}
$$

$$
RPAG_{l,gen,n} = \left\{ (gen', n') \in NBAG_{l,gen,n}; x(l, gen', n') > e_{repr} \right\} \tag{3.22}
$$

$$
\vartheta_{repr}^{gen,n}|X_{gen,n}(x)(x') =
$$

$$
\frac{1}{\#RPAG_{l,gen,n}} \sum_{(gen',n') \in RPAG_{l,gen,n}} \sum_{gen'' \in U}
$$

$$
mix(gen, gen')(gen'') \cdot cpchoose(x, x', gen, n, gen', n', gen'') \tag{3.23}
$$

$$
cpchoose(x, x', gen, n, gen', n', gen'') =
$$

$$
\begin{cases} [x' = x] \ if \ FC_{l,gen''} = \emptyset \\ \dfrac{1}{\#FC_{l,gen''}} \sum_{m \in FC_{l,gen''}} [x' = next(x, gen, n, gen', n', \\ gen'', m)] \ otherwise, \end{cases} \tag{3.24}
$$

where:

$$FC_{l,gen''} = \{o \in P_l \mid x(l, gen'', o) = 0\} \tag{3.25}$$

$$next(x, a, b, a', b', a'', m) = x_{next} :$$

$$x_{next}(i, j, k) = \begin{cases} x(i, j, k) - \frac{e_0}{2} & \text{if } j \in \{a, a'\} \text{ and } k \in \{b, b'\} \\ & \text{and } i = l \\ e_0 & \text{if } j = a'' \text{ and } k = m \text{ and } i = l \\ x(i, j, k) & \text{otherwise.} \end{cases} \tag{3.26}$$

Remark 3.2 ([51, Remark 6.1]) Note that if

$$\sum_{j=1}^{r} \sum_{k \in P_l} [x(l, j, k) > 0] < q_l,$$

which means that the location l is not full, then our assumptions guarantee that for every genotype *gen*

$$FC_{l,gen} \neq \emptyset,$$

i.e. there is a copy number to take.

Explanation The decision that the agent $ag_{gen,n}$ performs the action *repr*, i.e. $\delta_{repr}^{gen,n}$ (defined by Eq. (3.21)) is based on the condition that there is at least one neighboring agent in the same location and both agents have sufficient energy (higher than e_{repr}) to produce an offspring. This condition is verified by checking the contents of the set $RPAG_{l,gen,n}$ defined by Eq. (3.22). Therein $x \in X$ denotes the arbitrary state of the system when the decision is evaluated by the agent $ag_{gen,n}$.

In the case of positive evaluation of this decision (i.e., there exists $ag_{gen',n'}$ with sufficient energy) and enough space in the location (i.e., the number of agents does not exceed q_l), the state transition described by Eq. (3.23) is performed. This formula stems from Bayes' theorem which makes us check all possible agents from $RPAG_{l,gen,n}$. For each agent contained in this set a different state transition is performed, as described by function $next(\cdot, \cdot, \cdot, \cdot, \cdot, \cdot, \cdot)$. The probability of choosing an agent is equal to $(\#RPAG_{l,gen,n})^{-1}$.

The agent $ag_{gen,n}$ that initiated the action with its neighbor $ag_{gen',n'}$ becomes a parent and selects the offspring agent genotype gen'' using the probability distribution $mix(gen, gen') \in \mathcal{M}(U)$. The offspring agent ($ag_{gen'',n''}$) is created if there is enough room in the parental location (in this case there is always a free copy number, see Remark 3.2). If there is more than 1 inactive copy, the copy-number n'' of the offspring agent is selected with uniform probability distribution—see Eqs. (3.24) and (3.25).

The state transition function is constructed in the way described in Eq. (3.26). The incidence matrix $x_{next} \in X$ is obtained from x by changing entries related to agents $ag_{gen,n}, ag_{gen',n'}, ag_{gen'',n''}$. Part of the parents' energy is passed to the offspring agent with the genotype gen'', which is activated in the location l (and whose energy is set to e_0).

Observation 3.1.5 ([51, Observation 6.4]) *The value of the probability transition function imposed by Eqs. (3.10), (3.11), (3.21)–(3.26), performed by the agent $ag_{gen,n}$ present in the location l depends only on the elements of the system state contained in its location. This function does not introduce any changes in other locations.*

Explanation Equations (3.21)–(3.26) involve only those entries of the incidence matrix that are related to the location l, so both assumptions fully satisfy the conditions of Definition 3.1.

A decision on the migration action $migr$ may be undertaken by the agent with enough energy to migrate if there exists a location that is able to accept it (its number of agents does not exceed the maximum). When these conditions are met the agent is moved from one location to another one.

Observation 3.1.6 ([51, Observation 6.5]) *The probability transition function $\varrho_{migr}^{gen,n} : X \to \mathcal{M}(X)$ associated with the action migr is determined by:*

$$\delta_{migr}^{gen,n}|X_{gen,n}(x)(1) = \begin{cases} 1 \text{ if } (x(l, gen, n) > e_{migr} \text{ and } \#ACCLOC_l > 0) \\ 0 \text{ otherwise} \end{cases} \quad (3.27)$$

$$ACCLOC_l = \left\{ Loc \setminus \{l\} \ni l' : \big((l, l') \in Top\big) \right.$$

$$\left. \text{and } \left(\sum_{j=1}^{r} \sum_{k \in P_{l'}} [x(l', j, k) > 0] < q_{l'} \right) \right\} \quad (3.28)$$

$$\vartheta_{migr}^{gen,n}|X_{gen,n}(x)(x') =$$

$$\frac{1}{\#ACCLOC_l} \sum_{loc' \in ACCLOC_l} \frac{1}{\#FC_{loc',gen}}$$

$$\sum_{m \in FC_{loc',gen}} [x' = next(x, gen, n, loc', m)], \quad (3.29)$$

where $FC_{loc',gen}$ is given by Eq. (3.25) and

$$next(x, a, b, c, m) = x_{next} :$$

$$x_{next}(i, j, k) = \begin{cases} 0 & \text{if } i = l \text{ and } j = a \\ & \text{and } k = b \\ x(l, a, b) & \text{if } i = c \text{ and } j = a \\ & \text{and } k = m \\ x(i, j, k) & \text{otherwise.} \end{cases} \quad (3.30)$$

Explanation The decision on the action $migr$, $\delta_{migr}^{gen,n}$ (defined by Eq. (3.27)) is based on condition that the agent $ag_{gen,n}$ has sufficient energy to migrate (greater than $e_{migr} \geqslant e_{repr}$) and there is at least one neighboring location (l') capable of accepting the agent that wants to migrate (the number of agents in this location does not exceed $q_{l'}$), which is stated by checking the contents of the $ACCLOC_l$ set defined by Eq. (3.28). The arbitrary state of the system when the decision is evaluated by the agent $ag_{gen,n}$ is denoted by $x \in X$.

In the case of positive evaluation (of this decision), the state transition described by Eq. (3.29) is performed. Again, the formula comes from Bayes' theorem, which checks all possible locations from $ACCLOC_l$. For each location contained in this set, a different state transition is performed as described by the function $next(\cdot, \cdot, \cdot, \cdot)$. The probability of choosing a location is equal to $(\#ACCLOC_l)^{-1}$. The agent initiating the action moves from its location (l) to location l' that is uniformly chosen from the set $ACCLOC_l$. The change of location requires a change of the migrating agent's copy number. A new number is chosen from the set of available copy numbers for the target location (if the location is not full, this set is not empty, see Remark 3.2) according to the uniform distribution.

The state transition function is constructed as described in Eq. (3.30). The incidence matrix $x_{next} \in X$ is obtained from x by changing two entries related to a position of the agent $ag_{gen,n}$ in the location.

Observation 3.1.7 ([51, Observation 6.6]) *The value of the probability transition function imposed by Eqs. (3.10), (3.11), (3.27)–(3.30) and performed by the agent $ag_{j,k}$ present in the location l does not depend only on the elements of the system state contained in its location. The action $migr$ may introduce changes in the state entries associated with the location l and another location.*

Explanation Equations (3.27)–(3.30) include references to the elements of the system contained in the l-th and other locations.

The EMAS definition given here is enriched (with respect to the one in [50, 255]) by endowing it with a new cloning and mutation action clo, which allows a single agent to produce offspring.

A decision on the cloning and mutation action clo is based on checking the amount of agent's energy only. An agent with sufficient energy may create offspring agent based on its solution using the predefined family of probability distributions $mut : U \to \mathcal{M}(U)$, associated with genetic mutation (see, e.g. [290]). In particular, $mut(gen)(gen')$ denotes the probability that gen' is born of the parent gen. Part of the parent's energy ($e_1 = n_1 \cdot \Delta e, n_1 \in \mathbb{N}$) is passed onto the offspring agent.

By Observation 3.1.1 the following may be obtained:

Observation 3.1.8 ([51, Observation 6.7]) *The probability transition function $\varrho_{clo}^{gen,n} : X \to \mathcal{M}(X)$ associated with the action clo is determined by:*

$$\delta_{clo}^{gen,n} | X_{gen,n}(x)(1) = \begin{cases} 1 \text{ if } x(l, gen, n) > e_{repr} \\ \quad \text{and } \sum_{j=1}^{r} \sum_{k \in P_l} [x(l, j, k) > 0] < q_l \\ 0 \text{ otherwise} \end{cases} \qquad (3.31)$$

$$\vartheta_{clo}^{gen,n} | X_{gen,n}(x)(x') = \sum_{gen' \in U} mut(gen)(gen') \cdot cpchoose_1(x, x', gen, n, gen')$$

(3.32)

$$cpchoose_1(x, x', gen, n, gen') = \begin{cases} [x' = x] \text{ if } FC_{l,gen'} = \emptyset \\ \dfrac{1}{\#FC_{l,gen'}} \sum_{m \in FC_{l,gen'}} [x' = \\ next(x, gen, n, gen', m)] \text{ otherwise,} \end{cases}$$

(3.33)

where $FC_{l,gen'}$ is given by Eq. (3.25) and

$$next(x, a, b, a', m) = x_{next} :$$

$$x_{next}(i, j, k) = \begin{cases} x(i, j, k) - e_1 & \text{if } j = a \text{ and } k = b \\ & \text{and } i = l \\ e_1 & \text{if } j = a' \text{ and } k = m \\ & \text{and } i = l \\ x(i, j, k) & \text{otherwise.} \end{cases}$$

(3.34)

Explanation The decision $\delta_{clo}^{gen,n}$ on the action *clo* defined by Eq. (3.31) is based on condition that the energy of the agent exceeds the predefined threshold e_{clo}, where $x \in X$ denotes the arbitrary state of the system when the decision is evaluated by the agent $ag_{gen,n}$.

In the case of positive evaluation of this decision and enough space in the location (the number of agents does not exceed q_l), the state transition described by Eq. (3.32) is performed. As for the previous actions, this formula stems from Bayes' theorem. Based on the target agent (after applying a mutation operator), the state transition is performed as described by the function $next(\cdot, \cdot, \cdot, \cdot, \cdot)$.

The state transition function is constructed in the way described in Eq. (3.34). The incidence matrix $x_{next} \in X$ is obtained from x by changing entries related to agents $ag_{gen,n}$ and $ag_{gen',n'}$. Part of the parent's energy is passed onto the offspring agent with genotype gen' which is activated in location l and whose energy is set to e_0.

Observation 3.1.9 ([51, Observation 6.8]) *The value of the probability transition function imposed by Eqs. (3.10), (3.11), (3.31)–(3.34) and performed by the agent $ag_{gen,n}$ present in the location l, depends only on the elements of the system state contained in its location. The action clo will not introduce any changes in the other location.*

Explanation In Eqs. (3.31)–(3.34) entries of the incidence matrix not related to the current l-th location remain intact. Also the right-hand sides of these equations do not involve these entries.

Action Resulting From the Local Search Activation

Using a mechanism similar to the one included in the definition of action *clo*, it is possible to represent local searches invoked from particular points encoded by genotypes in U. The action which implements the local search will be called *lse*. The agent executing *lse* produces a new agent with a new genotype gen' which results from the application of the local search procedure *loc* starting from the parental genotype *gen*. The local search may be invoked by an agent with sufficient energy (greater than e_{repr}), thus the decision function $\delta_{lse}^{gen,n}|X_{gen,n}$ will have the same form as determined by Eq. (3.31).

The result of running the local method is characterized by the function $loc : U \rightarrow \mathcal{M}(U)$. In the case of stochastic local search (e.g. a strictly ascending random walk), the probability distribution $loc(gen)$ characterizes the result of running such method starting from the parental genotype gen. Of course, $loc(gen)$ need not (and in general will not) be strictly positive as it might be assumed in the case of the genetic mutation distribution $mut(gen)$. In the case of deterministic local method, $loc(gen)$ takes strictly one positive value for the genotype gen' obtained from applying this local method to gen. Of course, the loc function depends on both the local search algorithm and the fitness function corresponding to the optimization problem at hand.

Part $e_1 = n_1 \cdot \Delta e$, $n_1 \in \mathbb{N}$ of the parent's energy is passed to the offspring in much the same way as it is carried out during execution of the action *clo*. The above assumptions together with the Observation 3.1.1 lead to the following:

Observation 3.1.10 ([51, Observation 6.9]) *The probability transition function* $\varrho_{lse}^{gen,n} : X \rightarrow \mathcal{M}(X)$ *associated with the action lse is determined by the decision function* $\delta_{lse}^{gen,n}|X_{gen,n}$ *described by the Eq. (3.31) and the actions' kernel by the Eqs. (3.32)–(3.34) in which the function mut* $: U \rightarrow \mathcal{M}(U)$ *is replaced by the function* $loc : U \rightarrow \mathcal{M}(U)$.

The above observation might be verified in the same way as Observation 3.1.8 formulated and proved for the *clo* action. Similarly, without additional verification we may accept the following:

Observation 3.1.11 ([51, Observation 6.10]) *The value of the probability transition function* $\varrho_{lse}^{gen,n} : X \rightarrow \mathcal{M}(X)$ *imposed by the action lse executed by the agent* $ag_{gen,n}$ *present in the location l depends only on the elements of the system state contained in its location. The action lse will not introduce changes in other locations.*

Action's Taxonomy

Observations 3.1.3, 3.1.5, 3.1.7, 3.1.9 and 3.1.11 may be summarised in the following corollary:

Corollary 3.3 ([51, Corollary 6.1]) *The action migr is global, whereas the actions get, repr, clo and lse are local, i.e.*

$$Act_{loc} = \{get, repr, clo, lse\},$$
$$Act_{gl} = \{migr\}.$$

Observation 3.1.12 ([51, Observation 6.11]) *The probability transitions imposed by actions get, repr, migr, clo and lse satisfy the Markov condition (see, e.g. [28]).*

Explanation The probability transitions of the actions $\varrho_{get}^{gen,n}$, $\varrho_{repr}^{gen,n}$, $\varrho_{migr}^{gen,n}$, $\varrho_{clo}^{gen,n}$, $\varrho_{lse}^{gen,n}$ given by Eqs. (3.10), (3.11), (3.16)–(3.30) depend only on the current state $x \in X$ of the system.

3.1.3 EMAS Management

In order to obtain relaxed synchronization (i.e., agents present in locations may act in parallel), a dedicated timing mechanism must be introduced, which means that all state changes must be assigned to subsequent time moments $t_0, t_1, \ldots$ Now, the algorithmic description for CA will be considered, $LA_i, i \in Loc$ and MA presented in Pseudocodes 3.1.1, 3.1.2 and 3.1.3, respectively [51] constituting Hoare's rendezvous-like synchronization mechanism [145]. Note that here and later $\underline{a}(B)$ denotes the effect of randomly sampling one of the elements from the set B with the random distribution a; it is also assumed that the sets $localact, globalact \subset Act$ contain the local and global actions' signatures respectively.

In Fig. 3.1 the scheme of the synchronization mechanism is shown.

$CA = ag_{gen,n}$, present in the location i at every observable time moment chooses an action it wants to perform. It uses the probability distribution ω to choose from Act and asks its supervisor (LA_i) for permission, and using function $send()$, it sends a message with a chosen action. Then, it suspends its work and waits for permission (or denial) from LA_i using blocking function $b\_receive()$.

Both these functions are variadic. The first parameter in each function is always a target identifier, and the other parameters may be one or more values to be passed. In

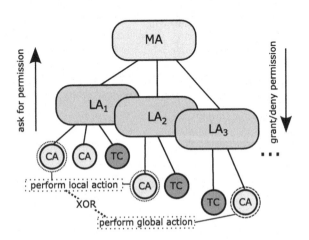

Fig. 3.1 EMAS
management structure

this particular case, the target either receives a certain value or just receives a signal from the sender (in this case no value is required).

Once permission is granted and the decision assigned to the action is true, the *CA* changes the state of the location (see Pseudocode 3.1.1). Then the agent suspends its work again in order to get permission to perform subsequent actions.

Pseudocode 3.1.1: COMPUTING AGENT'S ALGORITHM

while true
$$\begin{cases} reply \leftarrow 0 \\ \alpha \leftarrow \omega(x, gen)(Act) \\ send(LA_i, \alpha) \\ b\_receive(LA_i, reply) \\ \textbf{if } reply \textbf{ and } \underline{\delta_\alpha(x, gen, n)}(\{0, 1\}) \\ \quad \textbf{then } x_{next} \leftarrow \underline{\vartheta_\alpha^{gen,n}(x)}(X) \\ send(LA_i) \\ b\_receive(LA_i) \end{cases}$$

The LA_i (see Pseudocode 3.1.2) starts its work by checking whether the location contains any agents, so it sends a message to *MA* and waits for a reply. If there are any agents in the location, the LA_i receives signals containing identifiers of actions to be performed from all its agents and puts them into a hash map indexed by genotypes and containing actions identifiers. Then the LA_i utilizes the function $agsel_i$ to choose *CA* which should try to perform its action. This action is reported to *MA* and as soon as it has received permission, the computing agent can perform the action. All other agents (and the chosen one when permission is not granted) are stopped from performing their actions. Afterwards the *LA* waits for all *CA* to report the readiness to perform subsequent actions, and then reports this fact to the *MA*, and as soon as it has received permission, *CA* can perform their actions.

MA (see Pseudocode 3.1.3) waits for all requests from each location and then chooses randomly one location. If this location asks for permission to perform a global action, then this permission is granted and all other locations are rejected. Otherwise all locations asking for permission to perform a global action are rejected and all those asking for permission to perform local actions—are granted. Finally, *MA* waits once more for all locations to report that their work has been finished and let them try to perform a subsequent action.

3.2 Formal Analysis of EMAS

Following the concept of EMAS and its structural model provided in the previous section, discussion on asymptotic features of this system is given here. Starting with

Pseudocode 3.1.2: LOCAL AGENT'S ALGORITHM

while true
$\begin{cases} localgen \leftarrow \{(j, k) \in U \times P_i; \ x(i, j, k) > 0\} \\ genact \leftarrow hashmap(U \times P_i, Act) \\ act \leftarrow 0 \\ reply \leftarrow 0 \\ \textbf{if } \#localgen = 0 \\ \quad \textbf{then } \begin{cases} send(MA, null) \\ b\_receive(MA) \\ send(MA, null) \\ b\_receive(MA) \end{cases} \\ \quad \textbf{else } \begin{cases} \textbf{for each } g \in localgen \\ \quad \textbf{do } \begin{cases} b\_receive(g, act) \\ genact[g] \leftarrow act \end{cases} \\ gchosen \leftarrow agsel_i(x)(Act) \\ \text{REPORT } (genact[gchosen], gchosen) \end{cases} \end{cases}$

function REPORT($act, chosen$)
$send(MA, act)$
$b\_receive(MA, reply)$
if $reply$
  then $send(chosen, 1)$
  else $send(chosen, 0)$
for each $g \in (localgen \setminus chosen)$ **do** $send(g, 0)$
for each $g \in localgen$ **do** $b\_receive(g)$
$send(MA)$
$b\_receive(MA)$
for each $g \in localgen$ **do** $send(g)$

a detailed description of the system transition function, the ergodicity theorem is formulated and proved [51].

3.2.1   EMAS Dynamics

At every observable moment at which EMAS has the state $x \in X$, all CA in all locations notify their LA_i of their intent to perform an action. All LA_i choose an agent using the distribution given by the function $agsel_i(x)$ and then notify the Master Agent of their intent to let one of their agents perform an action. MA chooses the location with the probability distribution $locsel(x)$.

The probability that in the chosen location $i \in Loc$ the agent wants to perform a local action is as follows:

Pseudocode 3.1.3: MASTER AGENT'S ALGORITHM

```
while true
    ⎡ local ← {i : i ∈ [1, s]}
    │ localloc ← ∅
    │ localglob ← ∅
    │ act ← 0
    │ rep ← 0
    │ for each j ∈ local
    │         ⎡ b_receive(j, act)
    │      do │ if act ∈ Act_gl
    │         │    then localglob ← localglob ∪ {j}
    │         ⎣    else localloc ← localloc ∪ {j}
    │ lchosen ← locsel(x)(Loc)
    │ if lchosen ∈ localglob
    │            ⎡ send(lchosen, 1)
    │      then ⎨ for each j ∈ (local \ {lchosen})
    │            ⎣    do send(j, 0)
    │            ⎡ for each j ∈ localloc do send(j, 1)
    │      else ⎨ for each j ∈ localglob do send(j, 0)
    │ for each j ∈ local do b_receive(j)
    ⎣ for each j ∈ local do send(j)
```

$$\xi_i(x) = \sum_{gen \in U} \sum_{n \in P_i} agsel_i(x)(gen, n) \cdot w(x, gen)(Act_{loc}) \tag{3.35}$$

The probability that *MA* will choose the location with the agent which intends to perform a local action is:

$$\zeta^{loc}(x) = \sum_{i \in Loc} locsel(x)(i) \cdot \xi_i(x) \tag{3.36}$$

Of course, the probability that *MA* will choose the global action is:

$$\zeta^{gl}(x) = 1 - \zeta^{loc}(x) \tag{3.37}$$

If the global action has been chosen then the probability of passing from the state $x \in X$ to $x' \in X$ can be computed using Bayes rule as the sum over all possible sampling results:

$$\tau^{gl}(x)(x') = \sum_{i \in Loc} locsel(x)(i)$$

$$\left(\sum_{gen \in U} \sum_{n \in P_i} agsel_i(x)(gen, n) \cdot \left(\sum_{\alpha \in Act_{gl}} w(x, gen)(\alpha) \cdot \varrho_\alpha^{gen,n}(x)(x') \right) \right) \tag{3.38}$$

Now, a set of action sequences containing at least one local action is defined:

$$Act_{+1loc} = \left\{ (\alpha_1, \ldots, \alpha_s) \in Act^s; \ \sum_{i=1}^{s}[\alpha_i \in Act_{loc}] > 0 \right\} \qquad (3.39)$$

The following family of coefficients is defined $\{\mu_{\alpha_i, gen_i, n_i}(x)\}$, $i \in Loc$, $gen_i \in U$, $n_i \in P_i$, $x \in X$. If the location i is non-empty in the state x, then $\mu_{\alpha_i, gen_i, n_i}(x)$ is equal to the probability that the agent ag_{gen_i, n_i} residing in the i-th location will choose action α_i:

$$\mu_{\alpha_i, gen_i, n_i}(x) = agsel_i(x)(gen_i, n_i) \cdot \omega(x, gen_i)(\alpha_i). \qquad (3.40)$$

Of course, $\mu_{\alpha_i, gen_i, n_i}(x) = 0$ if the agent ag_{gen_i, n_i} does not exist in the location i in the state x, because $agsel_i(x)(gen_i, n_i) = 0$ in this case. Moreover, $\mu_{\alpha_i, gen_i, n_i}(x) = 1$, if the location i is empty in the state x. Next, the multi-index is introduced:

$$ind = \left(\alpha_1, \ldots, \alpha_s; (gen_1, n_1), \ldots, (gen_s, n_s) \right) \in IND = Act^s_{+1loc} \times \prod_{i=1}^{s}(U \times P_i).$$
$$(3.41)$$

The probability that in the state x, agents ag_{gen_i, n_i} will choose the actions α_i in consecutive locations is given by:

$$\mu_{ind}(x) = \prod_{i=1}^{s} \mu_{\alpha_i, gen_i, n_i}(x) \qquad (3.42)$$

Similarly to the previous case the probability of passing from the state $x \in X$ to $x' \in X$ for the parallel system can be computed using Bayes' rule as the sum over all possible sampling results:

$$\tau^{loc}(x)(x') = \sum_{ind \in IND} \mu_{ind}(x)(\pi_1^{ind} \circ, \ldots, \circ \pi_s^{ind})(x)(x'), \qquad (3.43)$$

where

$$\pi_i^{ind}(x) = \begin{cases} \varrho_{\alpha_i}^{gen_i, n_i}(x), & \alpha_i \in Act_{loc} \text{ and } i \text{ is non-empty} \\ \vartheta_{null}, & \alpha_i \in Act_{gl} \quad \text{or} \quad i \text{ is empty.} \end{cases} \qquad (3.44)$$

The definition of the coefficient $\mu_{\alpha_i, gen_i, n_i}(x)$ and the above formula (3.44) show in particular that the action *null* is executed in every location instead of a selected global action, and formally in all empty locations.

It is easy to see that the value of $(\pi_1^{ind} \circ, \ldots, \circ \pi_s^{ind})(x)(x')$ does not depend on the composition order because transition functions associated with local actions commute pairwise (see Corollary 3.3). It validates the following observation.

Observation 3.2.1 ([51, Observation 5.1]) *The probability transition function for the EMAS model is given by the formula*

$$\tau(x)(x') = \zeta^{gl}(x) \cdot \tau^{gl}(x)(x') + \zeta^{loc}(x) \cdot \tau^{loc}(x)(x') \tag{3.45}$$

and Eqs. (3.35)–(3.44).

Observation 3.2.2 ([51, Observation 5.2]) *The stochastic state transition of EMAS given by Eq. (3.45) satisfies the Markov condition. Moreover, the Markov chain defined by these functions is stationary.*

Proof All transition functions and probability distributions given by Eqs. (3.35)–(3.44) depend only on the current state of the system, which motivates the Markovian features of the transition function τ given by (3.45). The transition functions do not depend on the step number at which they are applied, which motivates the stationarity of the chain. □

3.2.2 Ergodicity of EMAS

It is intended to analyze some asymptotic features of the model in order to draw significant conclusions on capabilities of finding the optimum of a given function by EMAS with actions definitions given in Sect. 3.1.2.

The actions defined in Sect. 3.1.2, forming the following sets:

$$Act_{loc} = \{get, repr, clo, lse\},$$
$$Act_{gl} = \{migr\}.$$

are used in this proof.

Theorem 3.4 ([51, Theorem 7.1]) *Assume the following conditions hold.*

1. *The migration energy threshold is lower than the total energy divided by the number of locations $e_{migr} < \frac{1}{s}$. This assumption ensures that there will be at least one location in the system in which an agent is capable of performing migration (by gathering enough energy from its neighbors).*
2. *The quantum of energy is lower than or equal to the total energy divided by the maximal number of agents that may be present in the system $\Delta e \leqslant \frac{1}{\sum_{i=1}^{s} q_i}$. This assumption allows to achieve a maximal population of agents in the system.*
3. *The reproduction (cloning) energy is lower than two energy quanta $e_{repr} \leqslant 2\Delta e$.*
4. *The amount of energy passed from parent to child during the action clo is equal to Δe (so $n_1 = 1$).*
5. *The maximal number of agents on every location is greater than 1, $q_i > 1, i = 1, \ldots, s$.*
6. *Locations are connected to each other, i.e. $Top = Loc^2$.*

7. *Each active agent can be selected by its LA_i with strictly positive probability, i.e.*
   $\exists\ \iota_{agsel} > 0;\ \forall\ i \in Loc, \forall\ gen \in U,$
   $\forall n \in P_i,\ \forall\ x \in \{y \in X;\ y(i, gen, n) > 0\}, agsel_i(x)(gen, n) \geqslant \iota_{agsel}.$
8. *Families of probability distributions which are parameters of EMAS have uniform, strictly positive lower bounds*
   $\exists\ \iota_\omega > 0;\ \forall\ x \in X,\ gen \in U, \alpha \in Act, \omega(gen, x)(\alpha) \geqslant \iota_\omega,$

   $\exists\ \iota_{cmp} > 0;\ \forall\ gen, gen' \in U, cmp(gen, gen') \geqslant \iota_{cmp},$

   $\exists\ \iota_{mut} > 0;\ \forall gen, gen' \in U, mut(gen)(gen') \geqslant \iota_{mut},$

   $\exists\ 0 < \iota_{locsel} < 1;\ \forall\ x \in X, \forall\ j \in Loc,\ locsel(x)(j) \geqslant \iota_{locsel}.$

Then the Markov chain modeling EMAS (see Eq. (3.45)) is irreducible, i.e. the system state may be transformed between any two arbitrarily chosen states $x_b, x_e \in X$.

Remark 3.5 ([51, Remark 7.1]) Note that assumption 7 of Theorem 3.4 is reasonable because the number of possible states of the system is finite and so is the number of locations.

Remark 3.6 ([51, Remark 7.2]) The definition of the state space X (see Eq. (3.3)) implies that there already exists at least one computing agent in EMAS and that at least one location is non-empty at any time.

Because of the complexity, the technical details of the proof are transferred to Sect. 3.2.3.

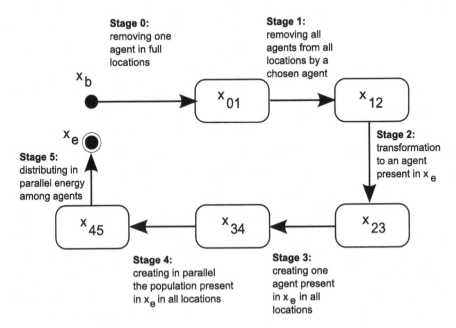

Fig. 3.2 State transitions in EMAS irreducibility proof

Proof of Theorem 3.4 It will suffice to show that the transformation between two arbitrary EMAS states $(x_b, x_e \in X)$ may be performed in a finite number of steps with the positive probability. The following sequence of stages (see Fig. 3.2) defines such transformation.

- **Stage 0**: In every location in parallel:
  - If the location is full, the agent is chosen and it performs sequentially the action *get* with one of its neighbors in order to remove it (to enable incoming migration from any other location). After removing one of its neighbors the agent tries to perform any global action, e.g. *migr* (and fails) until the end of the stage.
  - If the location is empty, the trivial null state transition is performed.
  - If the population in the location contains at least one agent, but is not full, this agent also attempts to perform the *migr* action (but fails to do it) during the whole stage. Final state of the *Stage 0* is denoted by x_{01}.

- **Stage 1**: A single location is chosen, in which the sum of agents' energy exceeds the migration threshold in the state x_{01} (based on Assumption 1 of Theorem 3.4 there must be at least one agent). Then, the agent ag_{gen_1,n_1} from this location (possibly possessing the largest energy in the state x_{0e}) is chosen. This agent performs a sequence of actions *get* in order to gather all energy from all its neighbors, finally removing them from the system (by bringing their energy to zero). Now ag_{gen_1,n_1} begins the first migration round in order to visit all locations and to remove the agents (overtaking their energy by performing multiple *get* actions). This round is finished in the location i_1. Now, the agent possesses the total energy of the system which equals 1. The final state of *Stage 1* is denoted by x_{12}. Note that the state matrix has only one positive entry $x_{12}(i_1, gen_1, n_1') = 1$ where $n_1' \in P_{i_1}$ is the new copy number of the selected agent after all migrations performed during the first round have occurred.

- **Stage 2**: The chosen agent $ag_{gen_1,n_1'}$ performs the *clo* action producing ag_{gen_2,n_2}, $n_2 \in P_{i_1}$, which is one of the agents present in the state in the location i_2. The location i_2 will contain the total energy greater than the migration threshold e_{migr} in the state x_e. Next, the agent $ag_{gen_1,n_1'}$ passes all its energy to this newly produced agent, finally being removed from the system. The purpose of *Stage 2* is to ensure that the agent recreating the population in the last location i_2 will be one of the agents present in this location in the state x_e.

- **Stage 3**: Next, the agent ag_{gen_2,n_2} begins the second migration round (starting migration from the location i_1), and visits all locations. In every visited location i it performs the *clo* operation producing one of the agents $ag_{gen_i^{first},n_i^{first}}$ that will be present in this location in the state x_e. In each non-empty location the cloned agent will receive total energy that should be assigned to its location in the state x_e by the sequence of *get* operations. The agent finishes migration in the location i_2 (one of the islands containing a total energy in the state x_e greater than the migration threshold e_{migr}). For the sake of simplicity the migrating agent after it reaches the final i_2 location will be further called in the same manner ($ag_{gen_i^{first},n_i^{first}}$).

- **Stage 4**: Every agent $ag_{gen_i^{first}, n_i^{first}}$ present in each non-empty location performs in parallel a sequence of cloning actions recreating the population of agents in its location in the state x_e. The total number of parallel steps is not greater than the maximal number of agents in the single location in the state x_e. Some agents may finish recreation earlier, and in this case they will choose the action *migr* (and fail to perform it) until the end of the stage.
- **Stage 5**: In every location in parallel: the agent active in *Stage 4* performs a sequence of *get* actions with its neighbors in order to pass them the a sufficient amount of energy required in the state x_e.

It was shown that each of the aforementioned stages requires performing at most a finite number of Markov chain steps. Moreover, it was shown that every aforementioned sequence has non-zero probability. For the details of the these features refer to Sect. 3.2.3 including detailed estimates of the lower probability bounds and upper bounds of the number of steps for every stage of the proof.

Remark 3.7 ([51, Remark 7.3]) Theorem 3.4 leads straightforwardly to the statement that each possible state of EMAS is reachable after performing a finite sequence of transitions independently of the initial population. Therefore, also the states containing the extrema are reachable. Thus any metaheuristic respecting EMAS architecture and the assumption of Theorem 3.4 satisfies the asymptotic guarantee of success [150, 241].

Theorem 3.8 ([51, Theorem 7.2]) *If the assumptions of Theorem 3.4 hold, then the Markov chain modeling EMAS is aperiodic.*

Proof Consider a state of the chain such that each location contains a single agent. In this case let us assume that each agent chooses *get* as its next action. Because all agents have chosen local actions, *MA* will allow them all to perform their actions, however, the absence of neighbors will force all the agents to perform the trivial (i.e. null) action. The transition probability function is then the s-fold composition of ϑ_{null}—see Eq. (3.43). Therefore, in this case the system will return to the same state in one step. The probability of such transition is not less than $(\iota_{get})^s > 0$. It means that the state is aperiodic. The chain is irreducible (see Theorem 3.4) and therefore has only one class of states, the whole state space, which obviously contains the aperiodic state. On the other hand, from Theorem 2.2 of [151] it is clear that aperiodicity is a state class property. In this case it means that all states of EMAS are aperiodic, which concludes the proof.   □

The following corollary is a consequence of Theorems 3.4 and 3.8.

Corollary 3.9 ([51, Corollary 7.1]) *The Markov chain modeling EMAS is* ergodic.

Remark 3.10 ([51, Remark 7.4]) The Markov chain (3.45) is ergodic in its strong sense, namely it is not only irreducible, but also aperiodic. Such chains are often called *regular* (see e.g. [151]). Obviously it is also ergodic in its weaker (and also wide) sense, meaning that it is simply irreducible.

Since the space of states X is finite we may introduce the probability transition matrix:

$$Q = \{\tau(x)(y)\}, \ x, y \in X, \tag{3.46}$$

where τ is the EMAS probability transition function—see Eq. (3.45). The Markov chain describing the EMAS dynamics is a sequence of random variables (or, equivalently, probability distributions) $\{\xi_t\} \subset \mathcal{M}(X), t = 0, 1, \ldots$ where ξ_0 should be a given initial probability distribution. Of course we have that

$$\xi_{t+1} = Q \cdot \xi_t, \ t = 0, 1, \ldots \tag{3.47}$$

Remark 3.11 ([51, Remark 7.5]) By Theorems 3.4 and 3.8 as well as the ergodic theorem [28] there exists a strictly positive limit $\widehat{\xi} \in \mathcal{M}(X)$ (i.e., $\widehat{\xi}(x) > 0, \forall x \in X$) of the sequence $\{\xi_t\}$ as $t \to +\infty$. This equilibrium distribution does not depend on the initial probability distribution ξ_0.

3.2.3 Technical Details of EMAS Ergodicity Proof

The detailed estimation of lower bound for probabilities and upper bound for number of steps for the stages of proof of Theorem 3.4 is preceded by a series of useful technical lemmas.

Lemma 3.12 ([51, Lemma B.1]) *Given the assumptions of Theorem 3.4, there exists a positive constant* $0 < \zeta_0 \leqslant \frac{1}{2}$ *such that* $\zeta_0 \leqslant \zeta^{gl}(x)$ *and* $\zeta_0 \leqslant \zeta^{loc}(x)$ *for all* $x \in X$.

Proof According to Formula (3.36)

$$\zeta^{loc}(x) = \sum_{i \in Loc} locsel(x)(i) \cdot \xi_i(x)$$

where

$$\xi_i(x) = \sum_{gen \in U} \sum_{n \in P_i} agsel_i(x)(gen, n) \cdot \omega(x, gen)(Act_{loc})$$

See Eq. (3.35). Because $Act_{loc} \neq \emptyset$ and there always exists at least one agent (see Remark 3.6) we have

$$\xi_i(x) \geqslant \iota_{agsel} \cdot \iota_\omega$$

for all $x \in X$ and $i = 1, \ldots, s$. Finally, because at least one location contains an agent (see Remark 3.6 again) we may evaluate

$$\zeta^{loc}(x) \geqslant \iota_{locsel} \cdot \iota_{agsel} \cdot \iota_\omega = \zeta_0$$

for all $x \in X$. Replacing Act_{loc} by Act_{gl} ($Act_{gl} \neq \emptyset$) we similarly obtain

$$\zeta^{gl}(x) \geq \iota_{locsel} \cdot \iota_{agsel} \cdot \iota_\omega = \zeta_0$$

for all $x \in X$. The constant ζ_0 is strictly positive as it is the product of strictly positive numbers. Moreover $2\zeta_0 \leq \zeta^{gl}(x) + \zeta^{loc}(x) = 1$ for all $x \in X$, so $\zeta_0 \leq \frac{1}{2}$.   □

Lemma 3.13 ([51, Lemma B.2]) $\forall gen \in U, \ l, l' \in Loc; l \neq l', \ n \in P_l, \ n' \in P_{l'}, x, x' \in X$ so that $x(l, gen, n) > e_{migr}, \ x(l', gen, n') = 0, x'(l', gen, n') = x(l, gen, n)$ we have

$$\varrho_{migr}^{gen,n}(x)(x') \geq \iota_{migr} = \frac{1}{(s-1) \max_{i \in Loc}\{q_i\}}.$$

In other words, assuming the current state x, and the particular agent $ag_{gen,n}$ ready to migrate from this state, all states x' containing this agent located in other locations than in x are reachable with probability greater than or equal to ι_{migr}.

Proof It follows straightforwardly from Eq. (3.29) describing the *migr* action's kernel $\varrho_{migr}^{gen,n}$.   □

Lemma 3.14 ([51, Lemma B.3]) *If $\exists \iota_{mut} > 0; \ mut(gen)(gen') \geq \iota_{mut} \ \forall gen, gen' \in U$ then $\exists \ \iota_{clo} > 0$ such that $\varrho_{clo}^{gen,n}(x)(x') \geq \iota_{clo}$ for each quadruple $(gen, n, x, x'), \ gen \in U, \ x, x' \in X$ and for a certain location $l \in Loc$ satisfying:*

- $x(l, gen, n) > e_{repr}, \ n \in P_l,$
- $\exists \ gen' \in U, n' \in P_l;$
  $x'(l, gen', n') = \Delta e, x(l, gen', n') = 0,$
- $\sum_{a \in U, b \in P_l}[x(l, a, b) > 0] < q_l.$

Roughly speaking, assuming the current state x, and the particular agent $ag_{gen,n}$ ready for cloning in this state, all states x' containing an additional agent (cloned by $ag_{gen,n}$) are reachable with probability greater than or equal to ι_{clo} independently upon x and x'. Of course, the set of possible x' may be empty if the location of the cloning agent is full in the state x.

Proof First of all, we may observe that if x satisfies the assumptions of the lemma, then the decision is positively evaluated, i.e. $\delta_{clo}^{gen,n}(x)(1) = 1$. On the other hand, the third assumption implies that there is at least one inactive agent in the system. Let us denote the signature of this agent by (gen', n'). If it was activated in the i-th location by the cloning operation, then the next state would satisfy $x'(l, gen', n') = \Delta e$ with probability 1. Then according to Eq. (3.25) the set $FC_{l,gen'}$ would be non-empty. Furthermore, taking into account (3.33), (3.32) we obtain

$$\varrho_{clo}^{gen,n}(x)(x') = \vartheta_{clo}^{gen,n}(x)(x') > \frac{\iota_{mut}}{\max_{i \in Loc}\{q_i\} - 1} = \iota_{clo} > 0.$$

    □

Lemma 3.15 ([51, Lemma B.4]) *If* $\exists \iota_{cmp} > 0$; $cmp(gen, gen') \geqslant \iota_{cmp}$
$\forall gen, gen' \in U$ *then* $\exists \iota_{get} > 0$ *such that* $\varrho_{get}^{gen,n}(x)(x') \geqslant \iota_{get}$ *and* $\varrho_{get}^{gen,n}(x)(x'') \geqslant \iota_{get}$
for each tuple (gen, n, x, x', x''), *$gen \in U$, $n \in P_l$, $x, x', x'' \in X$ and for a certain*
location $l \in Loc$ satisfying:

- $x(l, gen, n) > 0$, $n \in P_l$,
- $\exists \, gen' \in U, n' \in P_l$;
  $(gen, n) \neq (gen', n')$, $x(l, gen', n') > 0$,
- $x'(l, gen', n') = x(l, gen', n') + \Delta e$,
  $x'(l, gen, n) = x(l, gen, n) - \Delta e$,
  $x''(l, gen', n') = x(l, gen', n') - \Delta e$,
  $x''(l, gen, n) = x(l, gen, n) + \Delta e$,
  $x''(l, j, k) = x'(l, j, k) = x(l, j, k)$,
  $\forall j \neq gen, j \neq gen', k \neq n, k \neq n'$.

In other words, assuming the current state x, and a pair of agents $ag_{gen,n}$, $ag_{gen',n'}$
active in this state in the same location, both states x', x'' in which the agent $ag_{gen,n}$
takes or gives the quantum Δe of energy to/from its neighbor $ag_{gen',n'}$ are reachable
with probability greater than or equal to ι_{get}. This probability does not depend on
the state x or pair of the neighboring agents $ag_{gen,n}$, $ag_{gen',n'}$.

Proof First of all, we may observe that if x satisfies the conditions assumed in
the lemma then the decision is positively evaluated $\delta_{get}^{gen,n}(x)(1) = 1$, because
$NBAG_{l,gen,n}$ (see Eq. (3.17)) is non-empty. Then according to Eqs. (3.18)–(3.20)

$$\varrho_{get}^{gen,n}(x)(x') = \vartheta_{get}^{gen,n}(x)(x') > \frac{\iota_{cmp}}{(\max_{i \in Loc}\{q_i\}) - 1} = \iota_{get} > 0.$$

The same reasoning leads to $\varrho_{get}^{gen,n}(x)(x'') > \iota_{get}$. $\qquad\square$

Lemma 3.16 ([51, Lemma B.5]) *Let A_i be an event (e.g. denoting that certain agents*
perform certain actions) in the i-th step. Then $A_1 \cap \ldots \cap A_k$ is an event consisting of
events $A_1, \ldots, A_k$ taking place consecutively in subsequent steps $1, \ldots, k$. If $P(A_1) >$
$\lambda_1 > 0$ and the conditional probabilities $P(A_i| \cap_{j=1}^{i-1} A_j)$ are bounded from below by
$\lambda_i > 0$ for $i = 2, \ldots, k$, then

$$P\left(\bigcap_{i=1}^{k} A_i\right) \geqslant \prod_{i=1}^{k} \lambda_i > 0. \tag{3.48}$$

Proof Considering the sequence of (possibly dependent) events $A_1, \ldots, A_k$ and starting from the well-known conditional probability formula $P(A_1 \cap A_2) = P(A_1) \cdot$
$P(A_2|A_1)$, the following equation

$$P\left(\bigcap_{i=1}^{k} A_i\right) = P(A_1) \cdot \prod_{i=2}^{k} P\left(A_i \middle| \bigcap_{j=1}^{i-1} A_j\right) \tag{3.49}$$

may be proved inductively, which is enough to prove the lemma.   □

Proof of Theorem 3.4 – *detailed estimations* It will suffice to show that the passage between two arbitrary EMAS states (x_b, $x_e \in X$) may be performed in a finite number of steps with a positive probability. It was already shown (see *Outline od the proof of Theorem* 3.4, Sect. 3.2.2) that the passage mentioned above may be performed by the sequence of stages 1–5 described there and illustrated in Fig. 3.2. Now it is enough to estimate the upper bounds for the number of steps required to perform each of these stages and then the lower bound of the probability of series of actions executed in these steps.   □

Lemma 3.17 ([51, Lemma B.6]) *Given the assumptions of Theorem 3.4, Stage 0 requires at most* $st_0 = m - 1$ *steps in parallel taken with probability greater than or equal to* $pr_0 > 0$, *where m stands for the number of possible energy values that might be possessed by the agent (see Sect. 3.1).*

Proof In each parallel step performed during this stage we divide the set of locations into three distinct sets:

- $Loc_\emptyset$: empty locations (containing no active agents) – there is no activity there.
- Loc_{get}: locations containing the maximal number of agents (q_i): one of the existing agents ag_{gen_i,n_i} is selected and it performs a sequence of *get* actions in order to remove one of its neighbors $ag_{gen_i',n_i'}$. Both agents are fixed during the *Stage 0*. After removing its neighbour, the agent begins to perform the global action *migr* and fails (rejected to do it by *MA*), until the end of the *Stage 0*.
- $Loc_{migr} = Loc \setminus (Loc_\emptyset \cup Loc_{get})$: other locations in which one of the existing agents performs the global action *migr* and fails (rejected by *MA*), until the end of the Stage 0.

The probability of performing a single step of the described sequence is given by:

$$\zeta^{loc}(z) \prod_{i \in Loc_{get}} \left(agsel_i(z)(gen_i, n_i) \cdot \omega(gen_i, z)(get) \cdot \right.$$
$$\left. \delta_{get}^{gen_i,n_i}(z)(1) \cdot \varrho_{get}^{gen_i,n_i}(z_i')(z_i'') \right)$$
$$\prod_{i \in Loc_{migr}} \left(agsel_i(z)(gen_i, n_i) \cdot \omega(gen_i, z)(migr) \right)$$
$$\geqslant \zeta_0 \prod_{i \in Loc_{get}} \left(\iota_{agsel} \cdot \iota_\omega \cdot \iota_{get} \right) \cdot \prod_{i \in Loc_{migr}} \left(\iota_{agsel} \cdot \iota_\omega \right)$$
$$\geqslant \zeta_0 \left(\iota_{agsel} \cdot \iota_\omega \cdot \iota_{get} \right)^s \tag{3.50}$$

where:

- z is the current state of the system. It is set to x_b in the beginning of *Stage 0* and the final state of this stage z equals to x_{01},

- z_i', z_i'' are intermediate states that make it possible to express the parallel execution of the action get in all locations in Loc_{get}. The state z_i'' is a result of execution of the action get in the i-th location, starting from the state z_i'. The order of locations from Loc_{get} in Eq. (3.50) is arbitrary, because the get action is local (see Observation 3.1.3). $z_i' = z$ for the first location from Loc_{get}, $z_j' = z_i''$ where j is a location next to i in Eq. (3.50). The state z becomes z_i'' at the end of the sequence of intermediate steps.

It is easy to see that the estimation given in Eq. (3.50) is uniform for all steps in *Stage 0*. The maximal number of steps is bounded from above by $st_0 = m - 1$ because it is maximum possible energy for an agent.

Assume A_t is an event that consists in performing the t-th step described above, according to Lemma 3.16, the probability of the whole sequence can be bounded from below by:

$$pr_0 = \left(\zeta_0 \left(\iota_{agsel} \cdot \iota_w \cdot \iota_{get} \right)^s \right)^{m-1} > 0 \qquad (3.51)$$

$\square$

Lemma 3.18 ([51, Observation B.7]) *Given the assumptions of Theorem 3.4 and assuming Stage 0 to have been completed, Stage 1 requires at most $st_1 = s(m-1) + s - 1$ steps taken with probability greater than or equal to $pr_1 > 0$.*

Proof Following the assumptions of Theorem 3.4 there must be at least one location $i \in Loc$ with the total amount of energy greater than the migration threshold e_{migr} (of course at least one agent must be present there). An agent ag_{gen_1, n_1} is chosen from this location.

At each step of this stage the set of locations can be divided into three distinct sets:

- $Loc_\emptyset$: empty locations (containing no active agents)— there is no activity there.
- Loc_{get}: $\#Loc_{get} = 1$, single location containing the agent chosen at the beginning of the stage that performs a sequence of get actions in order to remove its neighbors (if they exist) in this location. Following the proof of Lemma 3.17 we may estimate the number of steps necessary to perform this sequence as $m - 1$.
- Loc_{migr}: other locations. The agents present in other non-empty locations are trying to perform the global action $migr$ but their requests are rejected by *MA*.

The probability of performing one action get in the sequence discussed is given by:

$$\zeta^{loc}(z)\bigg(agsel_i(z)(gen_1, n_\xi) \cdot \omega(gen_1, z)(get) \cdot \delta_{get}^{gen_1, n_\xi}(z)(1) \cdot \varrho_{get}^{gen_1, n_\xi}(z)(z')\bigg)$$

$$\cdot \prod_{j \in Loc_{migr}} \big(agsel_j(x)(gen_j, n_j) \cdot \omega(gen_j, z)(migr)\big)$$

$$\geqslant \zeta_0 \big(\iota_{agsel} \cdot \iota_\omega \cdot \iota_{get}\big) \cdot \prod_{i \in Loc_{migr}} \big(\iota_{agsel} \cdot \iota_\omega\big) \geqslant \zeta_0 \cdot \iota_{agsel} \cdot \iota_\omega \cdot \iota_{get} \qquad (3.52)$$

where z is the current state of the system. It is set to x_{01} at the beginning of *Stage 1* and the final state of this stage z equals x_{12}. The copy number of the chosen agent n_ξ is equal to n_1 at the beginning of *Stage 1* and then changes according to the migration rule becoming $n_1' \in P_{i_1}$ at the end of this stage. Note that the estimation given by Eq. (3.52) does not depend on the number of the step in the *Stage 1*. Then, the probability of removing all agents from a single location is given by:

$$\big(\zeta_0 \cdot \iota_{agsel} \cdot \iota_\omega \cdot \iota_{get}\big)^{m-1} \qquad (3.53)$$

After the removal of all its neighbors the chosen agent has to perform migration. The probability of this step is given by:

$$\zeta^{gl}(z) \cdot locsel(z)(i)$$

$$\bigg(agsel_i(z)(gen_1, n_\xi) \cdot \omega(gen_1, z)(migr)\delta_{migr}^{gen_1, n_\xi}(z)(1) \cdot \varrho_{migr}^{gen_1, n_\xi}(z)(z'')\bigg)$$

$$\geqslant \zeta_0 \cdot \iota_{locsel} \cdot \iota_{agsel} \cdot \iota_\omega \cdot \iota_{migr} \qquad (3.54)$$

where z'' is a state obtained after migrating of the chosen agent from one location to another.

Let us assume that A_t is an event that consists in removing all agents and migrating between the locations in the consecutive steps. The probability of each A_t may be evaluated by Eqs. (3.53)–(3.54). According to Lemma 3.16, the probability of the whole sequence can be bounded from below by:

$$pr_1 = \big(\zeta_0 \cdot \iota_{agsel} \cdot \iota_\omega \cdot \iota_{get}\big)^{s \cdot (m-1)} \big(\zeta_0 \cdot \iota_{locsel} \cdot \iota_{agsel} \cdot \iota_\omega \cdot \iota_{migr}\big)^{s-1} \qquad (3.55)$$

Moreover, the number of steps in the sequence may be estimated by the constant $st_1 = s(m-1) + s - 1$. $\qquad \square$

Lemma 3.19 ([51, Lemma B.8]) *Given the assumptions of Theorem 3.4 and assuming Stage 1 to have been completed, Stage 2 requires at most $st_2 = m$ steps taken with probability greater or equal to $pr_2 > 0$.*

Proof There is only one agent $ag_{gen_1, n_1'}$ in the system, so in order to produce another agent using *clo* it needs to perform 1 step. Then, it passes all its energy to its offspring by performing *get* action $(m - 1)$ times.

The probability of the first step of this sequence is as follows:

$$\zeta^{loc}(x_{12}) \cdot agsel_{i_1}(x_{12})(gen_1, n_1') \cdot \omega(gen_1, x_{12})(clo) \cdot$$
$$\delta_{clo}^{gen_1, n_1'}(x_{12})(1) \cdot \varrho_{clo}^{gen_1, n_1'}(x_{12})(z')$$
$$\geqslant \zeta_0 \cdot \iota_{agsel} \cdot \iota_\omega \cdot \iota_{clo} \tag{3.56}$$

where z' is the state where ag_{gen_2, n_2} was introduced into the system after performing the cloning action by $ag_{gen_1, n_1'}$.

Now after cloning, the agent $ag_{gen_1, n_1'}$ performs at most $(m-1)$ times get action to pass all its energy to ag_{gen_2, n_2}. A single step of this sequence has the following probability:

$$\zeta^{loc}(z)\left(agsel_{i_1}(z)(gen_1, n_1') \cdot \omega(gen_1, z)(get) \cdot \delta_{get}^{gen_1, n_1'}(1) \cdot \varrho_{get}^{gen_1, n_1'}(z)(z'')\right)$$
$$\geqslant \zeta_0 \cdot \iota_{agsel} \cdot \iota_\omega \cdot \iota_{get} \tag{3.57}$$

where z is the current state (at the end of the stage z will be equal to x_{23}) and z'' is the state after $ag_{gen_1, n_1'}$ passed a part of its energy to ag_{gen_2, n_2}.

Assuming A_t is an event that consists in performing the t-th step of the sequence described above, according to Lemma 3.16, the probability of *Stage 2* will be evaluated from below by:

$$pr_2 = \zeta_0 \cdot \iota_{agsel} \cdot \iota_\omega \cdot \iota_{clo} \cdot (\zeta_0 \cdot \iota_{agsel} \cdot \iota_\omega \cdot \iota_{get})^{m-1} > 0 \tag{3.58}$$

and $st_2 = 1 + (m-1) = m$. □

Lemma 3.20 ([51, Lemma B.9]) *Given the assumptions of Theorem 3.4 and assuming Stage 2 to have been completed, Stage 3 requires at most $st_3 = (m+1)^s$ steps taken with probability greater than or equal to $pr_3 > 0$.*

Proof Assuming *Stage 2* to have been completed, there is only one non-empty location i_1 containing agent ag_{gen_2, n_2}. The agent starts the second round of migration by performing the action $migr$. It is assumed that the path of this round is composed of a sequence of locations that ends at the i_2-th location. The location i_2 is one of the non-empty locations in the state x_e that contains total energy higher than the migration threshold e_{migr}. The length of this path is at most s. Note that the path is not intersecting, with the possible exception of the last location.

The probability of migration of the agent ag_{gen_2, n_2} between current location i to the next location of this path is given by:

$$\zeta^{gl}(z) \cdot locsel(z)(i)$$
$$\left(agsel_i(z)(gen_2, n_\xi) \cdot \omega(gen_2, z)(migr)\delta_{migr}^{gen_2, n_\xi}(z)(1) \cdot \varrho_{migr}^{gen_2, n_\xi}(z)(z')\right)$$
$$\geqslant \zeta_0 \cdot \iota_{locsel} \cdot \iota_{agsel} \cdot \iota_\omega \cdot \iota_{migr} \tag{3.59}$$

where z is the current state (in the beginning $z = x_{23}$ at the end of the stage z will be equal to x_{34}), z' is a state obtained after migrating the agent ag_{gen_2, n_ξ} from current location i to the next one along the mentioned path, $n_\xi \in P_i$ is the current copy number of the migrating agent.

$Loc_{migr} \subset Loc$ will denote a non-empty subset of locations which does not contain the current location in which $ag_{gen_i^{first}, n_i^{first}}$ is cloned or feeded by the life energy ($i \notin Loc_{migr}$). We assume that certain agents ag_{gen_j, n_j}, $j \in Loc_{migr}$ try to perform global action $migr$ and their requests are rejected by the MA.

The probability of each step in which the new agent $ag_{gen_i^{first}, n_i^{first}}$ is produced is evaluated by:

$$\zeta^{loc}(z) \cdot agsel_i(z)(gen_2, n_\xi)$$
$$\cdot \omega(gen_2, z)(clo) \cdot \delta_{clo}^{gen_2, n_\xi}(z)(1) \cdot \varrho_{clo}^{gen_2, n_\xi}(z)(z')$$
$$\prod_{j \in Loc_{migr}} \left(agsel_j(z)(gen_j, n_j) \cdot \omega(gen_j, z)(migr) \right)$$
$$\geq \zeta_0 \cdot \iota_{agsel} \cdot \iota_\omega \cdot \iota_{clo} \cdot (\iota_{agsel} \cdot \iota_\omega)^{s-1} \tag{3.60}$$

where z is the current state and z' is the state in which $ag_{gen_i^{first}, n_i^{first}}$ is created in the system and again $n_\xi \in P_i$ is the current number of copy of the migrating agent.

Now the agent passes the sufficient amount of energy (required to bring the total sum of energy of the i-th location to the value perceived in the system state x_e) to agent $ag_{gen_i^{first}, n_i^{first}}$ by performing at most $(m-1)$ times get. The probability of one get action is here as follows:

$$\zeta^{loc}(z) \left(agsel_i(z)(gen_2, n_\xi) \cdot \omega(gen_2, z)(get) \cdot \delta_{get}^{gen_2, n_\xi}(z)(1) \cdot \varrho_{get}^{gen_2, n_\xi}(z)(z'') \right)$$
$$\prod_{j \in Loc_{migr}} \left(agsel_j(z)(gen_j, n_j) \cdot \omega(gen_j, z)(migr) \right)$$
$$\geq \zeta_0 \cdot \iota_{agsel} \cdot \iota_\omega \cdot \iota_{get} \cdot (\iota_{agsel} \cdot \iota_\omega)^{s-1} \tag{3.61}$$

where z is the current state, z'' is the state where ag_{gen_2, n_ξ} passed a part of its energy to $ag_{gen_i^{first}, n_i^{first}}$ and n_ξ as in the previous Eq. (3.60).

Assuming A_t to be the consecutive t-th event described above, according to Lemma 3.16, the probability of the whole *Stage 3* will be bounded from below by:

$$pr_3 = \left(\zeta_0 \cdot \iota_{locsel} \cdot \iota_{agsel} \cdot \iota_\omega \cdot \iota_{migr} \cdot \zeta_0 \cdot \iota_{agsel} \cdot \iota_\omega \cdot \iota_{clo} \cdot (\iota_{agsel} \cdot \iota_\omega)^{s-1} \right)^{s-1}$$
$$\left(\zeta_0 \cdot \iota_{agsel} \cdot \iota_\omega \cdot \iota_{get} \cdot (\iota_{agsel} \cdot \iota_\omega)^{s-1} \right)^{m-1} > 0. \tag{3.62}$$

The number of steps required for performing the whole sequence is $st_3 = 2 \cdot (s - 1) + (m - 1)$.   □

Lemma 3.21 ([51, Lemma B.10]) *Given the assumptions of Theorem 3.4 and assuming Stage 3 to have been completed, Stage 4 requires at most $st_4 = \max_{i \in Loc}\{q_i\}$ steps taken with probability greater than or equal to $pr_4 > 0$.*

Proof We divide the set of locations into three distinct sets:

- Empty locations $Loc_\emptyset \subset Loc$ (containing no active agents): there is no activity there.
- Locations $Loc_{migr} \subset Loc$ in which one agent attempts to perform *migr* action and fails (these locations contain one agent in the final state x_e or have just finished recreation of the agents in the final state).
- Other locations $Loc_{clo} \subset Loc$ in which one agent $ag_{gen_i^{first}, n_i^{first}}$ performs a sequence of the *clo* actions in order to recreate the population of its neighbors (required in the state x_e).

The probability of this step may be evaluated by:

$$\zeta^{loc}(z) \prod_{i \in Loc_{clo}} \left(agsel_i(z)(gen_i^{first}, n_i^{first}) \cdot \omega(gen_i^{first}, z)(clo) \cdot \right.$$
$$\left. \delta_{clo}^{gen_i^{first}, n_i^{first}}(z)(1) \cdot \varrho_{clo}^{gen_i^{first}, n_i^{first}}(z')(z'') \right)$$
$$\prod_{i \in Loc_{migr}} \left(agsel_i(z)(gen_i, n_i) \cdot \omega(gen_i, z)(migr) \right)$$
$$\geqslant \zeta_0 \prod_{i \in Loc_{clo}} \left(\iota_{agsel} \cdot \iota_w \cdot \iota_{clo} \right) \prod_{i \in Loc_{migr}} \left(\iota_{agsel} \cdot \iota_w \right)$$
$$\geqslant \zeta_0 \left(\iota_{agsel} \cdot \iota_w \cdot \min\{\iota_{clo}, \iota_w\} \right)^s \tag{3.63}$$

where z is the current state, $z = x_{34}$ at the beginning and $z = x_{45}$ at the end of the stage, gen_i, n_i is the quasi signature of an arbitrary agent in the location $i \in Loc_{migr}$. Note that the whole sequence of the *clo* actions on one location may have length $\max_{i \in Loc}\{q_i\}$ in the worst case, so $st_4 = \max_{i \in Loc}\{q_i\}$.

Assuming A_t to be an event that consists in performing the t-th step of the sequence described above, according to Lemma 3.16, the probability of the whole sequence will be bounded from below by:

$$pr_4 = \left(\zeta_0 \left(\iota_{agsel} \cdot \iota_w \cdot \min\{\iota_{clo}, \iota_w\} \right)^s \right)^{\max_{i \in Loc}\{q_i\}} > 0. \tag{3.64}$$

$\square$

Lemma 3.22 ([51, Lemma B.11]) *Given the assumptions of Theorem 3.4, Stage 5 requires at most $st_5 = m - 1$ steps in parallel taken with probability greater than or equal to $pr_5 > 0$.*

Proof We divide the set of locations into three distinct sets:

- Empty locations $Loc_\emptyset \subset Loc$ (containing no active agents): there is no activity there.
- $Loc_{migr} \subset Loc$: in these locations, the agent tries to perform the global action $migr$ but its requests are rejected by MA (these locations contain one agent in the final state x_e or have just finished redistribution of the agents' energy in the final state).
- Other locations $Loc_{get} \subset Loc$: the agent containing the highest amount of energy ($ag_{gen_i^{first}, n_i^{first}}$) performs a sequence of get actions in order to pass the sufficient amount of energy to all its neighbors (required in the state x_e).

After each agent $ag_{gen_i^{first}, n_i^{first}}$ has finished distributing energy it starts to perform the global action $migr$ but its requests are rejected, and waits for other agents to finish their sequences. The locations containing exactly one agent behave in the same way. In the worst case there will be $(s - 1)$ agents performing this global action.

The probability of performing one of the get action of the sequence described above is given by:

$$
\zeta^{loc}(z) \prod_{i \in Loc_{get}} \left(agsel_i(z)(gen_i^{first}, n_i^{first}) \cdot \right.
$$
$$
\omega(gen_i^{first}, z)(get) \cdot
$$
$$
\left. \delta_{get}^{gen_i^{first}, n_i^{first}}(z)(1) \cdot \varrho_{get}^{gen_i^{first}, n_i^{first}}(z')(z'') \right)
$$
$$
\prod_{i \in Loc_{migr}} \left(agsel_i(z)(gen_i, n_i) \cdot \omega(gen_i, z)(migr) \right)
$$
$$
\geqslant \zeta_0 \prod_{i \in Loc_{get}} \left(\iota_{agsel} \cdot \iota_\omega \cdot \iota_{get} \right) \prod_{i \in Loc_{migr}} \left(\iota_{agsel} \cdot \iota_\omega \right)
$$
$$
\geqslant \zeta_0 \left(\iota_{agsel} \cdot \iota_\omega \cdot \min\{\iota_{get}, \iota_\omega\} \right)^s \tag{3.65}
$$

where z is the current state, $z = x_{45}$ at the beginning and $z = x_e$ at the end of this stage and gen_i, n_i is the quasi signature of an arbitrary agent present in the location $i \in Loc_{migr}$.

Assuming A_t to be an event that consists in performing the t-th step of the sequence described above, according to Lemma 3.16, the probability of the whole sequence will be bounded from below by:

$$
pr_5 = \left(\zeta_0 \left(\iota_{agsel} \cdot \iota_\omega \cdot \min\{\iota_{get}, \iota_\omega\} \right)^s \right)^{m-1} > 0. \tag{3.66}
$$

$\square$

To conclude the proof of Theorem 3.4 let us note that Lemmas 3.17–3.22 together state the fact that the total number of steps necessary for passing between states x_b and x_e is not greater than

$$st = \sum_{a=0}^{5} st_a < +\infty. \tag{3.67}$$

Let us recall that all actions taken in the consecutive stages $0-5$ are executable, assuming the completion of the previous stages. The probability of completing each stage $a = 1, \ldots, 5$ was bounded from below independently on the state imposed by the previous stage by $pr_a > 0$. Thus the following positive real number

$$pr = \prod_{a=0}^{5} pr_a > 0 \tag{3.68}$$

estimates from below the probability of passing from x_b to x_e. As the states were taken arbitrarily, we can show analogously that one can pass from x_e to x_b with a positive probability, which concludes the proof. □

3.3 Formal Perspective on iEMAS

In this section, a description of EMAS state is cited and extended by adding a matrix describing iEMAS state (following [52]).

In order to define the iEMAS state, the EMAS state, which is a crucial part of iEMAS, must be outlined first. Here, similarly to Sect. 3.1, a set of three-dimensional, incidence and energy matrices is introduced, as $\lambda \in \Lambda$ with s layers (corresponding to all locations) $\lambda(i) = \{\lambda(i, gen, n), \ gen \in U, \ n \in P_i\}, \ i \in Loc$. The layer $\lambda(i)$ will contain energies of agents in the i-th location. In other words, if $\lambda(i, gen, k) > 0$, it means that the k-th clone of the agent with the gene $gen \in U$ is active, its energy equals $\lambda(i, gen, k)$ and it is present in the i-th location.

Now, the following coherency conditions are given:

- $(\cdot, j, k)$-th column contains at most one value greater than zero, which means that the agent with k-th copy of j-th genotype may be present in only one location at a time, whereas other agents containing copies of j-th genotype may be present in other locations;
- entries incidence and energy matrices are non-negative $\lambda(i, j, k) \geq 0, \ \forall \ i = 1, \ldots, s, \ j = 1, \ldots, r, \ k = 1, \ldots, p$ and $\sum_{i=1}^{s} \sum_{j=1}^{r} \sum_{k=1}^{p} \lambda(i, j, k) = 1$, which means that the total energy contained of the whole system is constant, equal to 1;
- each layer $\lambda(i)$ has at most q_i values greater than zero, which denotes the maximal capacity of the i-th location and moreover, the quantum of energy Δe is less than or equal to the total energy divided by the maximal number of individuals that may be present in the system $\Delta e \leq \frac{1}{\sum_{i=1}^{s} q_i}$, which allows us to achieve maximal population of agents in the system;
- reasonable values of p should be greater than or equal to 1 and less than or equal to $\sum_{i=1}^{s} q_i$; it is assumed that $p = \sum_{i=1}^{s} q_i$, which assures that each configuration

of agents in locations is available, in respect of the total number of active agents $\sum_{i=1}^{s} q_i$; increasing p over this value does not enhance the descriptive power of the model;

- the maximal number of copies for each location $\#P_i$ should not be less than q_i, because a system state in which a particular location is filled with clones of one agent should be allowed; increasing $\#P_i$ over q_i is only a formal constraint relaxation, so finally it is assumed that $\#P_i = q_i$.

Putting all these conditions together, a set of three-dimensional incidence and energy matrices may be described in the following way, giving the EMAS part of the system state.

$$\Lambda = \left\{ ince \in \{0, \Delta e, 2 \cdot \Delta e, 3 \cdot \Delta e, \dots, m \cdot \Delta e\}^{s \cdot r \cdot p}, \Delta e \cdot m = 1, \right.$$

$$\sum_{i=1}^{s} \sum_{j=1}^{r} \sum_{k=1}^{p} x(i, j, k) = 1, \forall i = 1, \dots, s : \sum_{j=1}^{r} \sum_{k=1}^{p} [x(i, j, k) > 0] \leq q_i, \quad (3.69)$$

$$\forall i = 1, \dots, s, \ j = 1, \dots, r, \ k \notin P_i : \ x(i, j, k) = 0,$$

$$\left. \forall j = 1, \dots, r, \ k = 1, \dots, p : \ \sum_{i=1}^{s} [x(i, j, k) > 0] \leq 1 \right\}$$

where $[\cdot]$ denotes the value of the logical expression in parentheses.

3.3.1 iEMAS State

To continue considerations presented in, e.g. [50, 52, 255] iEMAS has a dynamic collection of lymphocytes that belong to the finite set Tc. Lymphocytes are unambiguously indexed by the genotypes from U, so that $\#Tc = \#U = r$. Lymphocytes have a similar structure to the computing agents defined in the previous paragraph, however, their actions are different (because their goals are different from the computing agents' goals) and their total energy does not have to be constant.

In addition to the above-given EMAS state describing the location and energy of agents, a set of matrices containing similar information for lymphocytes must be considered. Yet there is no need to ensure the constant total energy for lymphocytes. This additional set of lymphocyte incidence and energy matrices is described in the following way:

$$\Gamma = \left\{ tcince \in [0, \Delta e, \dots, n \cdot \Delta e]^{r \cdot s} : \forall i = 1, \dots, s \ \sum_{j=1}^{r} [tcince(i, j) > 0] \leq tcq_j \right.$$

$$\left. \text{and } \forall j = 1, \dots, r \ \sum_{i=1}^{s} [tcince(i, j) > 0] \leq 1 \right\}, \quad (3.70)$$

where $tcince(i, j)$ stands for energy of tc_j which is active in the location i. The integers tcq_j, $j = 1, \ldots, s$ stand for the maximal number of lymphocytes in particular locations. It is most convenient to assume $tcq_j = q_j$, $\forall j = 1, \ldots, s$.

The space of iEMAS states is defined as follows:

$$X = \Lambda \times \Gamma. \tag{3.71}$$

Structure and Behavior of iEMAS

iEMAS may be modelled as the following tuple:

$$\langle U, \{P_i\}_{i \in Loc}, Loc, Top, Ag, \{agsel_i\}_{i \in Loc}, locsel, \{LA_i\}_{i \in Loc}, MA, \omega, Act,$$
$$\{typesel_i\}_{i \in Loc}, \{tcsel_i\}_{i \in Loc}, Tc, Tcact\rangle, \tag{3.72}$$

where:

- *MA* (Master Agent) is used to synchronise the work of locations; it allows performing actions in particular locations. This agent is also used to introduce necessary synchronization into the system;
- $locsel : X \rightarrow \mathcal{M}(Loc)$ is a function used by *MA* to determine which location should be permitted to perform the next action,
- LA_i (Local Agent) is assigned to each location; it is used to synchronize the work of computing agents present in the location; LA_i chooses a Computing Agent (*CA*) or a T-Cell (*TC*) and lets it evaluate a decision and perform the action, at the same time asking permission from *MA* to perform this action;
- $agsel_i : X \rightarrow \mathcal{M}(U \times P_i)$ is a family of functions used by LA_i to select the agent that may perform the action, such that each location $i \in Loc$ has its own function $agsel_i$.
- $\omega : X \times U \rightarrow \mathcal{M}(Act)$ is a function used by agents to select actions from the set *Act*.
- *Act* is a predefined, finite set of actions.
- $typesel_i$ is a function used to select the type of agent in i-th location to interact with the system in the current step,
- $tcsel_i$ is used to choose a lymphocyte in the i-th location to interact with the system in the current step,
- φ is the decision function for lymphocytes, which chooses an action for them to perform,
- *Tcact* is a set of actions that may be performed by lymphocytes.

Similarly to the case of EMAS (see Sect. 3.1) the population of agents is initialized by means of the introductory sampling. This may be regarded as a one-time sampling from X according to a predefined probability distribution (possibly the uniform one) from $\mathcal{M}(X)$. Each agent starts its work immediately after being activated. At every observable moment only one agent present in each location gains the possibility of changing the state of the system by executing its action.

The function $typesel_i$ is used by LA_i to decide, whether CA or a TC should be chosen. Then, one of the functions $agsel_i$ or $tcsel_i$ is used to determine which agent (or lymphocyte) present in the i-th location will be the next one to interact with the system. After being chosen, the agent $ag_{gen,n}$ chooses one of the possible actions according to the probability distribution $\omega(x, gen)$. In the case of lymphocyte tc_{gen}, the probability distribution $\varphi(x, gen)$ is used.

It must be noted that the selection of action by all agents, which carry the same genotype gen in the same state x, is performed according to the same probability distribution $\omega(x, gen)$ and does not depend on the the genotype copy number n. In the simplest case, ω returns the uniform probability distribution over Act for all $(x, gen) \in X \times U$. Similarly to the case of a lymphocyte, φ returns the uniform probability distribution over $Tcact$ for all $(x, gen) \in X \times U$.

Next, the computing agent or lymphocyte applies to LA_i for the permission to perform this action. When the necessary permission is granted, the agent $ag_{gen,n}$ (or the lymphocyte tc_{gen}) performs the action after checking that a condition defined by formulas (3.6) and (3.73) has been fulfilled. If during the action an agent's or lymphocyte's energy is brought to 0, this agent suspends its work in the system (it becomes inactive).

MA manages the activities of LA_i and allows them to grant their agents permissions to carry out requested tasks. The detailed managing algorithm based on the rendezvous mechanism [145] is described in Sect. 3.3.2.

Denote by X_{gen} a subset of states in which there are active agents with the genotype $gen \in U$ or an active lymphocyte. Again, as the first step in defining the iEMAS dynamics, the EMAS part of the system must be addressed.

Each action $\alpha \in Act$ will be represented as the a of function families $\left(\{\delta_\alpha^{gen}\}_{gen \in U}, \{\vartheta_\alpha^{gen}\}_{gen \in U} \right)$. The functions

$$\delta_\alpha^{gen} : X \to \mathcal{M}(\{0, 1\}) \tag{3.73}$$

represent the decision to be taken: whether the action can be performed or not. The action α is performed with the probability $\delta_\alpha^{gen}(x)(1)$ by the agent $ag_{gen,n}$ at the state $x \in X$ and rejected with the probability $\delta_\alpha^{gen}(x)(0)$.

Next, the formula

$$\vartheta_\alpha^{gen} : X \to \mathcal{M}(X) \tag{3.74}$$

defines the non-deterministic state transition functions, so that ϑ_α^{gen} is caused by the execution of the action α by the agent $ag_{gen,n}$. The function is only invoked if the agent in active, therefore it is enough to define its restriction $\vartheta_\alpha^{gen}|X_{gen}$ and take an arbitrary value on $X \setminus X_{gen}$.

If any action is rejected, the trivial state transition

$$\vartheta_{null} : X \to \mathcal{M}(X) \tag{3.75}$$

such that for all $x \in X$

$$\vartheta_{null}(x)(x') = \begin{cases} 1 \text{ if } x = x' \\ 0 \text{ otherwise} \end{cases} \tag{3.76}$$

is performed.

The probability transition function for the action α performed by the agent carrying the genotype gen

$$\varrho_\alpha^{gen} : X \to \mathcal{M}(X) \tag{3.77}$$

is given by the formula

$$\varrho_\alpha^{gen}(x)(x') = \delta_\alpha^{gen}(x)(0) \cdot \vartheta_{null}(x)(x') \tag{3.78}$$
$$+\delta_\alpha^{gen}(x)(1) \cdot \vartheta_\alpha^{gen}(x)(x'),$$

where $x \in X$ denotes a current state and $x' \in X$ a consecutive state resulted from the conditional execution of α.

The function $typesel_i$ is introduced, to choose which type of agents will be able to perform the action:

$$typesel_i : X \to \mathcal{M}(\{0, 1\}). \tag{3.79}$$

When 0 is chosen, one of the agents is activated, when 1—the lymphocyte is activated.

The function $agsel_i$ that chooses an agent to be activated is similar as in the case of EMAS but it now depends in some way on the extended state from X defined by (3.71). Now a new function that will choose a lymphocyte to be activated is introduced:

$$tcsel_i : X \to \mathcal{M}(Tc). \tag{3.80}$$

The function ω which chooses the action for the active agent remains intact, though its domain changes (because of the new state definition, see (3.71)).

The function which chooses the action for the active lymphocyte is the following:

$$\varphi : U \times X \to \mathcal{M}(Tcact) \tag{3.81}$$

Each action $\alpha \in Tcact$ will be represented as a pair of function families $\left(\{\gamma_\alpha^{gen}\}_{gen \in U}, \{\kappa_\alpha^{gen}\}_{gen \in U}\right)$. The functions

$$\gamma_\alpha^{gen} : X \to \mathcal{M}(\{0, 1\}) \tag{3.82}$$

represent the decision to be taken: whether the action can be performed or not. The action α is performed with the probability $\gamma\alpha^{gen}(x)(1)$ by the lymphocyte tc_{gen} at the state $x \in X$, and rejected with the probability $\gamma_\alpha^{gen}(x)(0)$.

The following family of functions $\eta_\alpha^{gen} : X \to \mathcal{M}(X)$ will be used, where $gen \in U$, $\alpha \in Tcact$. Each of them expresses the probability transition imposed by the lymphocyte tc_{gen} that performs the action $\alpha \in Tcact$. They are given by the general formula:

$$\eta_\alpha^{gen}(x)(x') = \gamma_\alpha(gen, x)(\{0\}) \cdot \vartheta_{null}(x)(x') + \gamma_\alpha(gen, x)(\{1\}) \cdot \kappa_\alpha^{gen,n}(x)(x')$$
$$(3.83)$$

The agents' and lymphocytes' actions may be divided into two distinct types: global—they change the state of the system in two or more locations, so only one global action may be performed at a time, and local—they change the state of the system inside one location respecting only the state of local agents; only one local action for one location may be performed at a time.

Therefore the *Act* set is divided in the following way: $Act = Act_{gl} \cup Act_{loc}$ and $Tcact : Tcact = Tcact_{gl} \cup Tcact_{loc}$ accordingly. Speaking informally, local actions (elements of Act_{loc}, $Tcact_{loc}$) change only the entries of the layer $x(i)$ of the incidence and energy matrices if the location $i \in Loc$ contains the agent performing a certain action. Moreover, these actions do not depend on other layers of x. The action *null* is obviously "the most local one", because it does not change anything at all.

In the case of EMAS and iEMAS, actions such as evaluation or lymphocyte pattern matching may be perceived as local, whereas the action of migration is perceived as global. The above-stated conditions may be defined formally and may be used to prove commutativity of iEMAS (cf. [50, 255]), as in the case of EMAS in [51].

3.3.2 *iEMAS Management*

Similarly to the case of EMAS described in Sect. 3.1.3, in order to design a Markov model of the system with relaxed synchronization (i.e. so that agents present in different locations may act concurrently), a timing mechanism must be introduced, i.e. all state changes must be assigned to subsequent time moments $t_0, t_1, \ldots$ [52].

In Fig. 3.3, a scheme of the synchronization mechanism is presented constituting Hoare's rendezvous-like synchronization mechanism [145], similarly to EMAS.

Fig. 3.3 iEMAS management structure

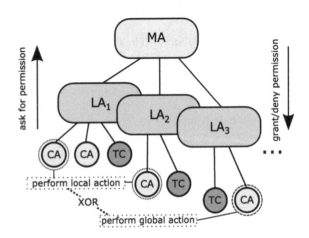

The *CA* (see Pseudocode 3.3.1) and *TC* (see Pseudocode 3.3.2) present in the location i at every observable time moment choose an action they want to perform and ask their supervisor (LA_i) for permission to carry on, sending a message with a chosen action identifier using function *send*() (similarly to the case of EMAS, see Sect. 3.1.3). Then they suspend their work and wait for permission (or denial) from LA_i using blocking function *b_receive*(). Both these functions are variadic. The first parameter in each function is always a target identifier, and the other parameters may be one or more values to be passed. In this particular case, the target either receives a certain value or just receives a signal from the sender (in this case no value is required).

Once the permission is granted and the decision assigned to the action is positive, the computational agent changes the state of the location. Then, the agent suspends its work again in order to get permission to perform a subsequent action.

LA_i (see Pseudocode 3.3.4) receives signals containing actions to be performed from all its agents. Then chooses one *CA* which should try to perform its action. This action is reported to *MA* and after receiving permission, the *CA* can perform the action. All other agents are stopped from performing their actions.

MA (see Pseudocode 3.3.3) waits for all requests from location and then chooses randomly one location. If this location asks for permission to perform a global action, then permission is granted this and all other locations are rejected. Otherwise all locations which asked for the permission to perform global action are rejected and all those asking for permission to perform local action are granted.

Pseudocode 3.3.1: COMPUTATIONAL AGENT'S ALGORITHM

while true
$$\left\{ \begin{array}{l} reply \leftarrow 0; \alpha \leftarrow \omega(x, gen)(Act) \\ send(LA_i, \alpha); b\_receive(LA_i, rep) \\ \textbf{if } rep \textbf{ and } \delta_\alpha^{gen}(x)(\{0, 1\}) \\ \quad \textbf{then } x_{next} \leftarrow \vartheta_\alpha^{gen}(x)(X) \\ send(LA_i); b\_receive(LA_i) \end{array} \right.$$

Pseudocode 3.3.2: LYMPHOCYTE'S ALGORITHM

while true
$$\left\{ \begin{array}{l} reply \leftarrow 0 \\ \alpha \leftarrow \varphi(x, tc_{gen})(Tcact) \\ send(LA_i, \alpha) \\ b\_receive(LA_i, rep) \\ \textbf{if } rep \textbf{ and } \delta_\alpha^{gen}(x)(\{0, 1\}) \\ \quad \textbf{then } x_{next} \leftarrow \eta_\alpha^{gen}(x)(X) \\ send(LA_i) \\ b\_receive(LA_i) \end{array} \right.$$

Pseudocode 3.3.3: MASTER AGENT'S ALGORITHM

while true
$\left[\begin{array}{l} local \leftarrow \{i : i \in [1, s]\} \\ localloc \leftarrow \emptyset \\ localglob \leftarrow \emptyset \\ act \leftarrow 0 \\ rep \leftarrow 0 \\ \textbf{for each } j \in local \\ \quad \textbf{do } \left\{\begin{array}{l} b\_receive(j, act) \\ \textbf{if } act \in \{Act_{gl} \cup Tcact_{gl}\} \\ \quad \textbf{then } localglob \leftarrow localglob \cup \{j\} \\ \quad \textbf{else } localloc \leftarrow localloc \cup \{j\} \end{array}\right. \\ lchosen \leftarrow \overline{locsel(x)(Loc)} \\ \textbf{if } lchosen \in \overline{localglob} \\ \quad \textbf{then } \left\{\begin{array}{l} send(lchosen, 1) \\ \textbf{for each } j \in (local \setminus \{lchosen\}) \textbf{ do } send(j, 0) \end{array}\right. \\ \quad \textbf{else } \left\{\begin{array}{l} \textbf{for each } j \in localloc \textbf{ do } send(j, 1) \\ \textbf{for each } j \in localglob \textbf{ do } send(j, 0) \end{array}\right. \\ \textbf{for each } j \in local \textbf{ do } b\_receive(j) \\ \textbf{for each } j \in local \textbf{ do } send(j) \end{array}\right.$

3.3.3 iEMAS Dynamics

Following the concept of iEMAS and its structural model provided in the previous section deliberations on asymptotic features of this system are given here. Starting with a detailed derivation of the system transition function, the ergodicity conjecture is formulated and its proof is outlined [52].

At every observable moment at which EMAS has the state $x \in X$ all agents in all locations notify their LA_i their intent to perform an action, all LA_i choose an agent with the distribution given by the function $agsel_i(x)$ and then notify the MA of their intent to let one of their agents to perform an action. MA chooses the location with the probability distribution $locsel(x)$.

The model of EMAS dynamics is extended here in order to model the behavior of iEMAS. The probability that in the chosen location $i \in Loc$ the agent or the lymphocyte wants to perform a local action is:

$$\xi_i(x) = typesel(x)(\{0\}) \sum_{gen \in U} \sum_{n=1}^{p} (agsel_i(x)(\{gen, n\})$$
$$\cdot \omega(gen, x)(Act_{loc})) + typesel(x)(\{1\}). \tag{3.84}$$

The probability that MA will chose the location with the agent intending to perform the local action is:

Pseudocode 3.3.4: *LA$_i$* ALGORITHM

while true

$\lceil$ *localgen* ← {(j, k) ∈ U × P$_i$; x(i, j, k) > 0}

$\mid$ *localtc* ← {U ∋ j : tcince(i, j) > 0}*genact* ← *hashmap*(U × P$_i$, Act)

$\mid$ *tcact* ← *hashmap*(U × P$_i$, Tcact)

$\mid$ *act* ← 0

$\mid$ *reply* ← 0

$\mid$ **if** #{*localgen* ∪ *localtc*} = 0

$\mid$ $\lceil$ *send*(MA, null)

$\mid$ **then** $\mid$ *b_receive*(MA)

$\mid$ $\mid$ *send*(MA, null)

$\mid$ $\lfloor$ *b_receive*(MA)

$\mid$ $\lceil$ **for each** g ∈ *localgen*

$\mid$ $\mid$ **do** $\lceil$ *b_receive*(g, act)

$\mid$ $\mid$ $\lfloor$ *genact*[g] ← act

$\mid$ $\mid$ **for each** g ∈ *localtc*

$\mid$ $\mid$ **do** $\lceil$ *b_receive*(g, act)

$\mid$ **else** $\mid$ $\lfloor$ *tcact*[g] ← act

$\mid$ $\mid$ **if** typesel(x)

$\mid$ $\mid$ **then** *gchosen* ← $\underline{agsel_i(x)}$(Act)

$\mid$ $\mid$ REPORT(*genact*[*gchosen*], *gchosen*)

$\mid$ $\mid$ **else** *gchosen* ← $\underline{tcsel_i(x)}$(Tcact)

$\lfloor$ $\lfloor$ REPORT(*tcact*[*gchosen*], *gchosen*)

function REPORT(*act*, *chosen*)

send(MA, act)

b_receive(MA, reply)

if *reply*

  then *send*(chosen, 1)

  else *send*(chosen, 0)

for each g ∈ (*localgen* \ *chosen*) **do** *send*(g, 0)

for each g ∈ *localgen* **do** *b_receive*(g)

send(MA)

b_receive(MA)

for each g ∈ *localgen* **do** *send*(g)

$$\zeta^{loc}(x) = \sum_{i \in Loc} locsel(x)(\{i\})\xi_i(x). \tag{3.85}$$

Of course, the probability that *MA* will choose the global action is:

$$(1 - \zeta^{loc}(x)) = \zeta^{gl}(x). \tag{3.86}$$

If the global action has been chosen, the state transition is given by:

$$\tau^{gl}(x)(x') = \sum_{i \in Loc} locsel(x)(\{i\}) \cdot \left(\sum_{gen \in U} \sum_{n=1}^{p} agsel(x)(\{gen, n\}) \cdot \right.$$

$$\left. \left(\sum_{\alpha \in Act_{gl}} \omega(gen, x)(\{\alpha\}) \cdot \varrho_\alpha^{gen,n}(x)(x') \right) \right). \tag{3.87}$$

The set of action sequences containing at least one local action is now defined:

$$Act_{+1loc} = \left\{ (\alpha_1, \dots, \alpha_s) \in (Act \cup Tcact)^s; \right.$$

$$\left. \sum_{i=1}^{s} [\alpha_i \in (Act_{loc} \cup Tcact)] > 0 \right\} \tag{3.88}$$

The probability that in the location i-th the agent ag_{gen_i,n_i} or the lymphocyte $tc_{\widetilde{gen_i}}$ will choose the action α_i is:

$$\mu_{\alpha_i,gen_i,n_i,\widetilde{gen_i}}(x) = typesel(x)(\{0\}) \cdot agsel_i(x)(\{gen_i, n_i\}) \omega(gen_i, x)(\{\alpha_i\}) +$$

$$typesel(x)(\{1\}) tcsel_i(x)(\{\widetilde{gen_i}\}) \varphi(\widetilde{gen_i}, x)(\{\alpha_i\}). \tag{3.89}$$

Now, a multi-index is defined:

$$ind = (\alpha_1, \dots, \alpha_s; (gen_1, n_1), \dots, (gen_s, n_s); (\widetilde{gen_1}), \dots, (\widetilde{gen_s}))$$

$$\in IND = (Act \cup Tcact)^s \times (U \times \{1, \dots, p\})^s \times U^s, \tag{3.90}$$

the probability that in consecutive locations agents ag_{gen_i,n_i} or lymphocytes $tc_{\widetilde{gen_i}}$ will choose the actions α_i is:

$$\mu_{ind}(x) = \prod_{i=1}^{s} \mu_{\alpha_i,gen_i,n_i,\widetilde{gen_i}}(x). \tag{3.91}$$

The transition function for parallel system is the following:

$$\tau^{loc}(x)(x') = \sum_{(\alpha_1,\dots,\alpha_s) \in Act_{+1loc}} \sum_{ind \in IND} \mu_{ind}(x) (\pi_1^{ind}(x) \circ, \dots, \circ \pi_s^{ind}(x))(x'), \tag{3.92}$$

where π_i is defined as:

$$\pi_i^{ind}(x) = \begin{cases} \varrho_{\alpha_i}^{gen_i,n_i}(x), & \alpha_i \in Act_{loc} \\ \eta_{\alpha_i}^{\widetilde{gen_i}}(x), & \alpha_i \in Tcact \\ \vartheta_{null}, & \alpha_i \in Act_{gl}. \end{cases} \tag{3.93}$$

The value of $(\pi_1^{ind}(x) \circ, \ldots, \circ \pi_s^{ind}(x))(x')$ does not depend on the composition order, because transition functions associated with local actions commutate pairwise (this feature of iEMAS actions may be proved similarly to the case of EMAS, cf. [51]). Finally, the following observation may be derived:

Observation 3.3.1 ([52, Observation 1]) *The probability transition function for the parallel iEMAS model is given by the formula*

$$\tau(x)(x') = \zeta^{gl}(x)\tau^{gl}(x)(x') + \zeta^{loc}(x)\tau^{loc}(x)(x') \qquad (3.94)$$

and formulas (3.84)–(3.93).

Observation 3.3.2 ([52, Observation 2]) *The stochastic state transition of iEMAS given by formula (3.94) satisfies the Markov condition.*

Proof All transition functions and probability distributions given by formulas (3.84)–(3.93) depend only on the current state of the system, which motivates the Markovian features of the transition function τ given by (3.94). The transition functions do not depend on the step number at which they are applied, which motivates the stationarity of the chain. □

3.3.4 Ergodicity of iEMAS

In this section, an ergodic conjecture of the Markov chain describing the behavior of iEMAS is presented. A sequence of proof stages is also given, it is shown that after estimating the upper bounds of their steps and lower bounds for their probabilities may become a full formal proof of iEMAS ergodicity in much the same way as it was realized for EMAS [52].

The system which uses the following actions (they may be defined in details in similar way as it is shown in Sect. 3.1.2), is considered in this section:

- *repr, clo, lse, migr* – these actions are inherited unchanged from EMAS.
- *get* – this action is modified. When the agent performs this action, and its energy (or energy of evaluated agent) reaches zero, it activates the lymphocyte containing the genotype of the inactivated agent.
- *give* – this is an action executed solely by lymphocytes. It is performed every time the lymphocyte is activated and decreases the lymphocyte's energy, which makes the lymphocyte be deactivated (when its energy reaches zero).
- *kill* – another lymphocyte's action, removing (or penalizing) the computing agent (performed when the genotype of the tested agent matches the pattern contained in the lymphocyte).

Features of these actions, in particular if they are global or local, may be proved similarly as in Sect. 3.1.2, resulting in the following actions taxonomy:

$$Act_{loc} = \{get, repr, clo, lse, give, kill\},$$
$$Act_{gl} = \{migr\}.$$

Conjecture 3.23 ([52, Theorem 1]) *Assume that the following assumptions hold.*

1. *The migration energy threshold is lower than the total energy divided by the number of locations $e_{migr} < \frac{1}{s}$. This assumption ensures that there will be at least one location in the system in which an agent is capable of performing migration (by gathering enough energy from its neighbors).*
2. *The quantum of energy is lower than or equal to the total energy divided by the maximal number of agents that may be present in the system $\Delta e \leqslant \frac{1}{\sum_{i=1}^{s} q_i}$. This assumption makes it possible to achieve a maximal population of agents in the system.*
3. *Reproduction (cloning) energy is lower than two energy quanta $e_{repr} \leqslant 2\Delta e$.*
4. *The amount of energy passed from parent to child during a cloning action is equal to Δe (so $n_1 = 1$).*
5. *The maximal number of agents in every location is greater than one, $q_i > 1, i = 1, \ldots, s$.*
6. *Locations are all connected, i.e. $Top = Loc^2$.*
7. *Each active agent can be selected by its LA_i with strictly positive probability.*
8. *The families of probability distributions which are the parameters of EMAS have uniform, strictly positive lower bounds.*

Then the Markov chain modeling iEMAS (see Eq. (3.94)) is irreducible, i.e. all its states communicate.

Proof outline 3.3.1 ([52, Sect. 7]) *In order to prove Conjecture 3.23, it is enough to show that the passage from x_b to x_e (two arbitrarily chosen states from X) may be performed in a finite number of steps with the probability strictly greater than zero (see Fig. 3.4).*

Consider the following sequence of stages [52].

- *Stage 0: In every location in parallel: If the location is full, an agent is chosen, and it performs sequentially an evaluation action with one of its neighbors in order to remove it (to make incoming migration possible from any other location, in case this location is full). After removing one of its neighbors the agent tries to perform any global action, e.g. migration (and fails), until the end of the stage. Otherwise, the trivial null state transition is performed. Final state of the Stage 0 is denoted by x_{0e}.*
- *Stage 1a: One location is chosen, at which the sum of agents' energy exceeds the migration threshold in the state x_{0e} (based on assumption 1 of Theorem 3.23 there must be at least one). Then one agent from this location ag_{gen_1,n_1} (possibly with the largest energy in the state x_{0e}) is chosen. This agent performs a sequence of evaluation actions in order to gather all energy from all its neighbors, finally removing them from the system (by bringing their energy to zero).*

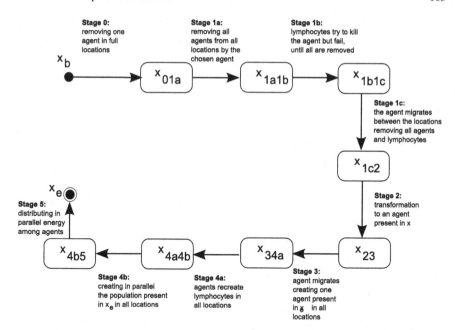

Fig. 3.4 State transitions in iEMAS irreducibility proof outline

- **Stage 1b**: *If there are any lymphocytes in the current location, they perform killing action, one by one, on the agent ag_{gen_1,n_1}, but fails to remove it from the system until all lymphocytes are removed. In the end, only one agent is present in the location.*

- **Stage 1c**: *Now this agent begins the first migration round in order to visit all locations and remove the agents (overtaking their energy by performing multiple get actions) and all lymphocytes. This round is finished at the location i_1. Now, agent ag_{gen_1,n_1} possesses the total energy of the system which equals 1. Final state of Stage 1 is denoted by x_{1e}. Note that the state matrix has only one positive entry $x_{1e}(i_1, gen_1, n_1) = 1$.*

- **Stage 2**: *The agent performs a cloning action producing one of the agents (ag_{gen_2,n_2}) that will be present in the location i_2, which is one of the locations in the state x_e containing the total energy greater than the migration threshold. Having passed all of its energy to this new agent, it is finally removed from the system. The purpose of Stage 2 is to ensure that the agent recreating the population in last location i_2 will be one of the agents present in this location in the state x_e. Otherwise, if i_2 is full in the state x_e, ag_{gen_1,n_1} cannot recreate this population. If ag_{gen_1,n_1} is active in the location i_2 at the state x_e (i.e. $x_e(i_2, gen_1, n_1) > 0$), Stage 2 may be omitted (in this case ag_{gen_1,n_1} takes the role of ag_{gen_2,n_2} in the consecutive stages).*

- **Stage 3**: *Next, the agent ag_{gen_2,n_2} begins the second migration round (starting migration from the location i_1) and visits all locations. In each visited location it performs a cloning action and produces one of the agents that will be present in this* *location in the state* x_e. *The cloned agent in each non-empty location (denoted by* $ag_{gen_i^{first}, n_i^{first}}$*) will receive the total energy that should be assigned to its location, by the sequence of evaluation actions. The agent finishes migration after recreating the population in the location* i_2 *(one of the islands containing the total energy in the state* x_e *greater than the migration threshold).*

- *Stage 4a: In the system, the following sequence of actions assigned with consecutive locations labeled* $i \in Loc$, *non-empty in the state* x_e, *is performed: each agent* $ag_{gen_i^{first}, n_i^{first}}$ *performs a cloning action to produce an agent with a genotype of one of the lymphocytes existing in the location in the state* x_e. *Now it performs a sequence of evaluation actions to remove the created agent, so the desired lymphocyte is produced. The lymphocyte performs a sequence of energy lowering actions to adjust its energy to the level observed in the state* x_e. *This is repeated until all the lymphocytes present in* x_e *are recreated.*
- *Stage 4b: In the system, the following sequence of actions assigned with the consecutive locations labeled* $i \in Loc$, *non-empty in the state* x_e, *is performed: every agent* $ag_{gen_i^{first}, n_i^{first}}$ *performs a sequence of cloning actions, recreating the population of agents in its location in the state* x_e.
- *Stage 5: In every location in parallel: agent* $ag_{gen_i^{first}, n_i^{first}}$ *performs a sequence of evaluation actions with its neighbors in order to pass to them a sufficient amount of energy, required in the state* x_e.

Assuming that Conjecture 3.23 is true, similarly to the EMAS case, the following features may be proved, which lead directly to the statement that every possible state of iEMAS is reachable (with positive probability) after performing a finite sequence of transitions independently of the initial population.

1. All states containing the extrema are reachable from an arbitrary initial state. Thus iEMAS satisfies asymptotic guarantee of success in the sense of [150, 241, 253] [52, Corollary 1].
2. If the assumptions of Theorem 3.23 hold, then the Markov chain modeling EMAS is aperiodic [52, Theorem 2].
3. As a consequence of features 1 and 2, the Markov chain modeling EMAS is *ergodic* [52, Corollary 2].
4. It is noteworthy that the Markov chain (3.94) is ergodic in its strong sense (not only irreducible but also aperiodic). Such chains are quite often called *regular* (see e.g. [151]) [52, Remark 1].
5. Since the space of states X is finite the probability transition matrix is introduced:

$$Q = \{\tau(x)(y)\}, \ x, y \in X, \tag{3.95}$$

where τ is the iEMAS probability transition function (see Eq. (3.94)). The Markov chain describing the iEMAS dynamics is a sequence of random variables (or, equivalently, probability distributions) $\{\xi_t\} \subset \mathcal{M}(X), t = 0, 1, \ldots$, where ξ_0 should be a given initial probability distribution. Of course, the following condition holds:

$$\xi_{t+1} = Q \cdot \xi_t, \ t = 0, 1, \ldots \tag{3.96}$$

6. By Theorems 3.23 and feature 4, as well as the ergodic theorem [28] there exists a strictly positive limit $\widehat{\xi} \in \mathcal{M}(X)$ (i.e., $\widehat{\xi}(x) > 0, \forall x \in X$) of the sequence $\{\xi_t\}$ as $t \to +\infty$. This equilibrium distribution does not depend on the initial probability distribution ξ_0 [52, Remark 2].

3.4 Summary

The theoretical results obtained in this research, especially the most important feature of ergodicity proved for EMAS (including its memetic versions) and iEMAS, are crucial for studying features of stochastic global optimization metaheuristics.

The strong ergodicity of the finite state Markov chain modeling the metaheuristics shows that these systems can reach an arbitrary state (arbitrary population) in the finite number of iteration with the probability equal to 1 which implies the asymptotic stochastic guarantee of success (see Remark 3.7).

The formal framework constructed made it possible to analyze similar systems (e.g. HGS [254] or parallel version of Vose's algorithm [256]) and will be considered for future analysis of other novel and classical computing systems (e.g. in the near future, an analysis of evolutionary algorithm using tournament selection is envisaged).

It is worth noting that apart from formal models focused on particular aspects of population-based metaheuristics (mostly Simple Genetic Algorithm [290, 293] and selected Evolution Strategies [247, 249]), there are no similar approaches to modeling such complex computing systems, as agent-based ones. It seems to be one of the greatest advantages of the research presented in this monograph.

Part II
Design and Implementation

Chapter 4
Agents and Components in Software Engineering

Agents introduce a conceptual model for distributed problem solving, declaring autonomy as a core feature of its building blocks. This makes them aware of the environment, support interactions among them and the environment and realize different actions focused on fulfilling the goal assigned the by the user or designer. It seems that implementation of such complex systems should follow an acclaimed methodology and needs adequate tools, in order to create robust, flexible and reliable software that will be easily extensible for all the interested users.

Component technologies provide an implementation model which seems to be a good method of choice for implementing complex systems as the abstraction of components can be easily translated into different elements possibly provided by different vendors. This allows for preparing frameworks that can be further parametrized and extended into particular systems. As one may notice agents and components have little in common at first sight, apart from being building blocks of software systems, the first in logical modelling, the second in implementation modelling.

This chapter focuses on presentation of agent-based and component paradigm leading to finding common subset of features that will allow to analyze and design the agent-based computing solutions described in the next chapter.

4.1 Technological Solutions for Agent-Based Systems

Because of vast importance of acquiring, exchanging and managing the information in the wide area network environment, the most typical field for the agent-based technology became heterogeneous distributed information systems. As a consequence, the most notable development of the agent-based technology was caused by the need to realize the mechanisms of sharing and discovering the services by the agents.

Contrary to the typical service-oriented architectures the communication interfaces are defined for agents in the intentional categories, creating new quality in

© Springer International Publishing AG 2017
A. Byrski and M. Kisiel-Dorohinicki, *Evolutionary Multi-Agent Systems*,
Studies in Computational Intelligence 680, DOI 10.1007/978-3-319-51388-1_4

the means of the tasks realized and increasing the possibilities of application [20]. Currently there are many advanced technologies making possible implementation of flexible communication mechanisms for cooperating, distributed agents, and accessing by them directory-related information. The key concept here is creating of proper standards and appropriate tools supporting construction of such systems and their cooperation in open network environment.

The concept of "agency" can bring interesting features in the applications related to simulation (cf. [283]), in particular in the case of high diversity of the problems (sociological, biological and others) connected with complex processes observed in populations consisting of a large number of different entities. Very often, macro-scale models are built for such problems, using appropriate mathematical methods that help in understanding the dynamics leading to perceiving different phenomena.

As an alternative, a micro-scale model may be considered, mimicking behaviors of cooperating entities and making possible observing of the phenomena arising on the level of the whole population. Such systems can be easily realized using agent-based approach. Moreover, it is easy to see, that implementation of such systems requires using of dedicated tools and frameworks, in particular different to these used for development of information systems.

4.1.1 Agent Platforms

Integration of the information that is distributed in the agent-based system (e.g. consolidation of the outcomes of the agents' actions) consists in assuming the existence of appropriate infrastructure—communication protocols and components responsible for collecting and providing the information about services delivered by the particular agents. These requirements are partially realized by the existing technological solutions (so-called *agent platforms*), delivering communication mechanisms and basic functions for identification and localization of agents and their services. Thus agent platform may be perceived as a particular kind of *middleware* supporting actions and communications among the agents as well as sometimes delivering functionalities of an integrated development environment supporting design, implementation and running of the agent-based systems.

The platform is usually distributed as a library of classes or modules, consisting of basic, easily extensible implementations of key functions of the system. The agent platforms can be very diverse considering the goal and the way of realization of particular agent features. Because of that, these platforms are very different considering their structure, agent model and the communication methods. Below the two popular agent-based platforms are described: JADE and MadKit.

JADE

JADE (Java Agent DEvelopment) is a Java-based environment facilitating implementation of distributed systems conforming to FIPA standards [18, 19]. FIPA,[1] Foundation for Intelligent Physical Agents, is an institution active since 1996 as non-profit international organization and currently a standardization committee of IEEE Computer Society. The goal of FIPA is working out the standards for agent-oriented heterogeneous information systems and one of its basic tasks is publishing of specifications considering cooperation (so-called interoperability, meant as common sharing and using the services) in the agent systems built using different tools.

FIPA specifications define among others basic assumptions for the system architecture, basic services delivered by the platform, agent communication method, interaction protocols and typical applications, focusing on external behavior of the system components, leaving unimplemented details of their inner structures and behavior. Communication in FIPA is based on theory of speech acts and consists in passing the messages in the form of communicative acts or performatives, describing context in which its contents should be interpreted, e.g. inform, request, refuse. The message is contained in a kind of envelope containing all the information required for understanding of its contents, in a form of agent communication language—ACL. Agent Message Transport (AMT) focuses on representation and the way of delivering the messages in different network environments. On the AMT level the message consists of an envelope, describing specific requirements and information required by the Message Transport Service—MTS on the agent platform, and encoded or compressed contents. Support for MTS on the agent platform is in the form of so-called *Agent Communication Channel* (ACC), providing communication method among the agents inside and outside the platform.

Agent Management (AM) provides a work environment for the agents—it fixes the logical reference model for creating, registering, localization, communication and migration of agents, describing the behavior of the Agent Platform (AP). It describes necessary platform elements and defines basic services as well as ontologies needed for their realization.

- *Agent Management System* (AMS)—naming, agent localization and authentication services,
- optional *Directory Facilitator* (DF)—registration and service localization services,
- previously described transport services (MTS).

JADE introduces two important organization entities:

- platform—can be distributed among several nodes, consists of many agent containers, the communication among them is realized using IIOP,
- agent container—a process that is transparent for agents, provides them with a complete execution environment and makes possible their running in parallel; the communication among the containers is realized using RMI.

---

[1]http://www.fipa.org.

Among the containers, there is a distinguished one, realizing mainly platform services and representing this platform for the outside clients. It provides implementation of the basic agents: AMS, DF and ACC (described in the previous section) that are activated during the initialization of the platform.

Each JADE agent is autonomous, i.e. it is not only reactive to the external stimuli, but also can act on its own (as it is implemented as a thread). The platform uses so-called *behaviors* for modeling of tasks, that later will be realized by the agents—in particular moment the agent can perform according to only one behavior.

The communication among the agents is realized using asynchronous exchange of messages. The agents have their own queues where the messages coming from other agents are placed (after placing the message the recipient is notified by the platform). The platform describes three main schemes of the communication:

- if the recipient is placed in the same container, the message is passed as an event, no unnecessary conversions are realized,
- if the recipient is placed in another container on the same platform, the message is passed using RMI,
- if the recipient is placed on another platform, ACC (Agent Communication Channel) is requested to deliver an appropriate messaging service. Next a relevant MTP (Message Transport Protocol) is found, that will be used for delivering the message.

Messages exchanged by the agents have particular structure described by ACL.

Madkit

MadKit (Multi-Agent Development Kit) is an agent-oriented platform based on AGR model (Agent—Group—Role) [139]. The agents belong to the groups for which particular roles are defined for accordingly acting agents. The agent can belong to one or more groups and act according to many roles. The role is an abstract representation of agent functions, service or recognition in a group. The group may be created by any agent, the agent must also require enrolling to any group (though this request can be declined).

MadKit structure is based on three following rules:

- Micro-kernel—the library has small, "lightweight" kernel, realizing the following tasks:
  - management of local groups and roles, management of the information about the group members and their roles, checking the correctness of requests concerning the groups and the roles,
  - management of agent life-cycles, the kernel creates and removes the agents, it is the only module that has direct references to agents, it also assures that the agent receives an unique identifier,
  - local message passing among the agents.

The kernel itself is also an agent, so the management and monitoring of its actions is also realized in a conformance to the agent model.

- "Agentification" of services—all services besides the ones provided by the micro-kernel are supported by the agents, e.g. the remote communication is supported by a dedicated communication agent.
- Component-oriented graphical user interface model.

The kernel is fully extensible using the *hooks* mechanism and kernel-focused operations, available to each agent belonging to the *system* group. Hooks realize the publish-and-subscribe model in the two following ways:

- *monitor hooks*—each operation realized by the kernel causes sending a particular information to the agents that have subscribed to this operation, thanks to this mechanism it is possible to monitor the agents and their structure,
- *interceptor hooks*— the particular operation is not realized, instead the agent is notified (the one that subscribed this operation), so changing the behavior of the basic functions of the platform becomes possible.

The agents communicate among themselves using asynchronous message exchange. The agent may send a message to another agent using a predefined address or to all agents with particular role in a group. The information is placed in a FIFO queue owned by the recipient agent.

The basic agent implementation is based on threads, i.e. each agent becomes an application thread. However it is also possible to create agents that are not threads.

MadKit makes possible development of distributed agent-based systems, while the fact of distribution is transparent to the user, i.e. he is not required to consider low-level communication aspects. In order to make a MadKit-based application a distributed one, no code modification is needed.

The platform is delivered as a set of plugins. Such structure makes possible adding, removing or modification of the agents and their features according to particular requirements.

4.1.2 Agent-Based Simulation Tools

In the agent-based simulation, a problem of important matter is synchronization of the agents working in parallel. In the simulations consisting of hundreds or millions of agents, classic problems of parallel problems (deadlock or starvation avoiding [5]) pose significant challenges. Such problem scale makes the implementation of agents as individual threads technologically difficult. A much better solution becomes the *inversion of control* approach, by using discrete event driven simulation (see, e.g. [243]), in particular, so called phase simulation. In this approach a dedicated mechanism for synchronization of agent group is defined, that makes possible for each agent to realize its action separately (querying the system state, changing of its own state) and later submit potentially conflicting actions, that will be executed later by the simulation manager. In this way the parallel programming of agents becomes a kind of *Round Robin* technique, making possible to perceive the particular simulation entities as autonomous and pseudo-parallel acting individuals, in a simulated way.

There exists many simulation environments that may be used for supporting of modeling and development of agent-based simulations. Some of them are oriented on particular kinds of simulation [223, 237], but there are also general-purpose tools. The most notable ones are: Swarm [214], RePast [225] and Mason [194], but also such tools as Galatea [70], SystemC [26], or SimPy[2] should also be mentioned.

Swarm

Swarm is a set of libraries that can be used for developing of agent-based models and simulations. Its fundamental organization entity is a *swarm* [214]. It consists of a collection of agents and their event schedule (concerning other agents). The simulation is realized in a discrete time steps. Each agents displaying certain behaviors and having its own state acts only when it is called by the schedule object or by other agents. Swarm makes possible building of hierarchical structures, where agent may consist of swarms of other agents. In this way the behavior of the supervising agent may be closely connected to the behavior of the agents belonging to its swarm.

Architecture of the Swarm library is based on two main swarms:

- model—the internal swarm consists of agents and conducts the simulation,
- observer—the external swarm consists of agents-observers which main goal is to monitor and manage the simulation using the mechanism of probes. They can e.g. present the data in the real-time or record them for further analysis.

Such dividing of the actual model from the "observation" code makes possible easy modification of these both parts of the system, independently on the other one. The mechanism of probes is a base for agent "observations". It can be used for dynamic acquiring of information from the particular agents or their groups. The probe connects to an agent, it can send a message as well as acquire and read its properties.

A serious disadvantage of Swarm library is lack of distribution mechanisms. The agents working in one process cannot communicate with the agents working in another one.

MASON

MASON is an agent-based simulation environment developed at the Georg Mason University. Its name not only is derived from the university name, but also can be read as an acronym: *Multi-Agent Simulator Of Neighbourhoods (or Networks)* [194]. Multi-layered architecture makes possible achieving full independence between the simulation logic and visualization tools. The models are fully independent and can be embedded into other Java-based programs.

---

[2]http://simpy.sourceforge.net/.

Simulation in MASON consists of a model class (`SimState`), that consists of a random number generator and (`Schedule`). The object of this class manages many agents by the `Steppable` interface. Each agent is instantiated as an object and can cooperate with other agents and with the environment using a dedicated method, called by the schedule (discrete event-driven simulation). The `SimState` class can also manage the spatial structure of simulation by the so-called (`Field`)s, which can be assigned with different objects, thus making possible the localization of the agents.

MASON is advertised as an efficient and portable environment (100% Java) that arises from its simplicity. There are no predefined communication mechanisms nor a system organization—these can be realized using simple method calls. There are also no embedded support for distributed simulations nor solutions concentrated on extensibility and easy configuration of the models.

RePast

RePast *Recursive Porous Agent Simulation Toolkit* is an advanced environment supporting development and execution of agent-based simulations [225]. There are several version of this platform available, e.g. Java, .NET or Python based, and the newest version (Repast Simphony) supports RAD approach making possible graphical programming of the application. There are RePast libraries available for different analysis and modeling methods, e.g. genetic algorithms, neural networks, social networks, GIS and others.

The simulation structure in RePast is similar to the one described above for MASON. The simulation objects undergo discrete simulation managed by a schedule using dedicated method calls. The environment fully supports parallel event execution, there is also HPC version available [64]. A very important feature is an introspection mechanism available in the GUI—in this way particular parameters of the simulation can be altered without the need for restart.

Similar to MASON, RePast does not have any embedded communication mechanism, besides the schedule event one. It does not provide any support for creating extensible model configurations.

4.2 Component-Based Technologies

Because of high complexity of today's software systems, a particularly important aspect of their development is the possibility of reusing their functional fragments. It makes possible not only the reduction of production costs, but also increasing the clarity and ease of complexity management of existing solutions. Because of that in the last years component or service-oriented application architectures constantly gain popularity, leading to the possibility of independent creation and reuse of parts of the systems [206].

Origins of the component-oriented approach to the software development should be sought mainly in object-oriented design and development [276]. Object-oriented paradigm introduces a concept of cooperating objects (by sharing and using services), makes possible creating abstract logic of the program, in a way reminding the analysis of the real-world phenomena (so-called object-oriented analysis). At the same time unification of two programming aspects (well known even in von Neumann architecture), namely data and instructions in one being is realized (so-called *encapsulation*). Another "root" of the component technologies is modular approach that consists in putting together the system using independently developed elements (modules, libraries). Though this approach assumes information hiding and low coupling, they are realized in static dependencies during the implementation phase [231].

The notion of "software component" was used for the first time in 1968 [208], and since this time, many different technologies were developed, declaring the support for the component orientation [276]. Among important examples one should notice System Object Model (SOM) developed by IBM, Component Object Model (COM) developed by Microsoft and CORBA standard proposed by Object Management Group. Most of these technologies imposed certain technological requirements, making the products incompatible with other environments, thus contradicting the theory of component portability. The ever increasing need for software tool making possible implementation of small and lightweight solutions connected to the concept of components leading to leveraging one of the object-oriented design rule, the inversion of control—*IoC*.

4.2.1 Components, Dependencies, Contracts

The simplest definition of a component was proposed by Clemens Szyperski in the introduction to his book [276], stating simply that *components are for composition*. Later he undertakes a complex analysis of different aspects of components and composition, starting from the production (implementation), through the mechanisms of distribution and acquisition and finishing with the problems of installation and deployment.

His classic definition assumes describes the following features of a component:

- it is an unit of independent development, i.e. it is well-separated from its environment and from other components;
- it is an unit of independent deployment, so it must have clear specification of its requirements and delivered services, so it may be integrated with other components,
- it does not have a persistent state, i.e. the component cannot be distinguished from its copies, so there is no sense of having more than one copy of the component.

However the above characteristics is not reflected in diversity of the component systems in the software market. Lack of standards making possible preparing a consistent characteristics of available solutions and overusing of terminology causes creation of divergent concepts and component models. As an example the notion of component presented in UML specification concentrates on its modular nature [226], completely ignoring all other aspects pointed out by Szyperski or other authors [206].

The relation between the capabilities of components and its requirements is usually expressed using a declarative specification of an *interface*, which plays the role of "access points" to the other components, making possible leveraging of the services shared by them [66]. Concrete implementations of the components delivering certain realizations of the interface that sometimes introduce higher dependencies on other components (e.g. realization of inheritance between the classes delivered by two different components). Such connections should not be realized in order to avoid the issues such as *fragile base class problem* [276]). Dependency relations are a criterion of reaching an agreement (a contract) between the components in a target composition environment, that authorizes their collaboration in the frames of one application. It is to note, that the contract may be interpreted in a context of a dynamics of the component deployment process in the target platform, as in the context of declarative specification on the level of the delivered configuration.

From the object-oriented point of view, component technologies are based on the following three important rules [202]:

- *Don't Repeat Yourself*—DRY is the basis of all reuse techniques and assumes, that each software element should have an unambiguous representation in the system model, that directly maps to the specification of the component dependencies.
- *Open-Closed Principle*—OCP claims, that the software elements (classes, modules, components etc.) should be open for extension and closed for modifications. Thanks to specification of the dependencies on the interface level, the components can be extended by additional functions, at the same time they cannot be modified behind the interfaces.
- *Dependency Inversion Principle*—DIP claims definition of the dependencies based on "abstractions" instead of realization details and forbids the dependency of the higher-level modules on the lower-level ones [201]. In the context of component-based approach it should be directly mapped onto specifications and realization of dependency mechanisms. The components should be dependent only on the abstractions in the form of interfaces and thus be able to cooperate with any entities declaring the appropriate interface.

Moreover, the last rule maps on the abstract notion of contract, that expresses the relation between the installed components, however the actual implementation of this mechanisms is to be realized by the component environment.

Dependent on the technology, contract realization mechanisms may be implemented using different approaches, leaving often the issue of the specification interpretation to the components (as e.g. in the case of *extension points* in the Eclipse RCP environment [205]). Moreover, when the installed components instantiate groups of connected objects, the mechanism of contract realization must deliver references to the required objects, that is realized using either *Service Locator* pattern, or the above-mentioned inversion of control rule.

4.2.2 Inversion of Control and Dependency Injection

The inversion of control paradigm in general consists in inversion of control of program logic flow, i.e. declaration of the needs instead of their active realization. This concept was first announced in 1988, when it was used for showing the flexibility of the architecture based on the software framework, overtaking the control over coordination of the program execution sequences [157]. The inversion of control rule is ofter described by a "Hollywood rule", *Don't call us; we'll call you*, that was proposed in a paper devoted to a high-level programming language Mesa designed for description of the dynamics of user interface [275]. The inversion of control rule affects to a great extent particular aspects of the organization of the system architecture and is nowadays very popular (e.g. in event-driven programming paradigm). However this rule is not defined precisely in the literature, and nowadays is inseparably connected with the *dependency injection*—DI mechanism.

Dependency injection is a design pattern, in which the inversion of control can be found in relation to the process of solving the dependencies, thus making possible to separate this process (*separation of concerns*—SoC) from the aspects of the object behavior logic, leading to decreasing of coupling between the classes. It means practically that the responsibility for the class instance creation is delegated to the external mechanism realizing the dependencies, so called IoC container. The most popular methods of delivering the dependencies is *Constructor injection, Setter injection* or *Interface injection*. Another kind of this mechanism, that is commonly used in the Java component environments is annotated *Field injection*.

From the perspective of the system architecture, *IoC container* is a central point of the application, that is responsible i.a. for realization of the process of object instantiation, managing the object life cycle and delivering fully configured instances on the user's demand. Definitions of the parts of the application along with specification of the required dependencies are passed to the IoC container in the form of configuration, that is dependent on particular technology. Most of the implementations of the IoC containers supports automatic dependency solving—so called auto-wiring, based on the names, attribute types, method signatures or annotations, that reduces or even eliminates the need for its specification. A very important functionality of the IoC container is management of so-called *scopes*, that may be interpreted as availability of the instances in the particular application. The most popular two scopes are *prototype* (new instances are created based on the definition independently on the context and the environment state) and *singleton* (in one IoC container there is only one instance of the created object, that is available in the particular context).

4.2.3 Component Environments

Independently on the accepted terminology, the component as an abstraction entity with specified dependencies must be deployed in an execution environment equipped with component framework, aware of the semantics of the delivered information and necessary mechanisms supporting composition and running of the system. The framework is fully responsible for management of the composition process, that consists in discovering and loading the components, possibly also realization of the life cycle.

Diversity of the products present in the software market makes possible undertaking a free choice of the technology for the needs of design of particular application (classic solutions were already mentioned in the introduction to this section). However at the same time this diversity hampers the reusability of the implemented components. A possible answer to this problem may be found in so-called *lightweight* component frameworks[3] which functionality is based on simple mechanisms, in particular the dependency injection pattern, thus making possible adaptation of the components from different systems. As examples of such techniques the following frameworks may be referenced: Pico Container,[4] Google Guice[5] and Spring Framework.[6] Unfortunately such programming tools concentrate mostly on solving the dependencies, do not cover many other problems connected with composition of independently-developed components, such as information sharing or versioning.

OSGi acronym stands for *Open Services Gateway Initiative* and describes an organization, specification and a software product. The base for all of the technologies developed in the field of OSGi initiative is a series of standarized specifications. The key element of the software based on *OSGi Service Platform* is *OSGi Framework*, that delivers a generic and fully manageable module system dedicated to Java language [206]. A composition unit here is a *bundle*, identified as a software module that can be run and stopped, updated and uninstalled in runtime, installed in the remote environment and individually managed. The functions that make the OSGi distinguishable among other similar solutions are advanced mechanisms for information hiding and module versioning making the process of evolution of the software components much easier.

From the point of view of the dependency management, the basic functionalities of the OSGi environment are exporting and importing the Java packages and importing the OSGi bundles. These mechanisms however lack dynamic behavior and the possibility of contract specification between the components using interfaces, that are provided by the so-called service layer. The interface allows for establishing a contract between the software modules, though without necessity of *ad hoc* delivering the implementation, and its realization is carried out by the selected OSGi packages, is treated as a service. Additional mechanisms of *Declarative Services* replaces the manual initialization of services by a declarative one (XML file). In the context of the already presented solutions, a very interesting feature of the OSGi is

---

[3]Literature does not introduce a detailed description of *lightweight* and *heavyweight* component technologies, these notions should rather be treated as expressions of certain trends in their structure and mechanism used by them.

[4]http://picocontainer.org/.

[5]http://code.google.com/p/google-guice/.

[6]http://www.springsource.org/.

the specification of *Blueprint Container*, defining a dependency injection container that is responsible for realization of the IoC paradigm at the same time using dynamic services of OSGi.

4.3   Summary

First publications about agents and agent-based systems were originally connected with the field of artificial intelligence, but as the time passed, this topic evolved in the technological direction, encompassing also the issues of development of typical systems of this class. The agents are applied in an increasing number of different problems, starting with small applications, like personal e-mail filters, and finishing with big systems dealing with important issues such as supporting the clinical treatment or flight control. Today it can be said that *agent-based systems* became an independent research field lying on the border of artificial intelligence and software engineering, and enormous popularity of agent-based approach is caused mainly by the simplicity of the description of a plethora of different complex system in the agent-based categories. It leads however to overusing of the agent notion and applying it by a "brute force" to describing any possible systems [301], that might be a side-effect of a significant scientific background of this paradigm.

Exactly opposite trend can be seen in the case of the component technologies. Their evolution lead today to a very broad diversity both in technology and terminology connected with different environments and application fields. It is connected with difficulties of integration of different software solutions that is reflected with the products available on the market. A significant change of this state can be predicted only if there are acclaimed generally available mechanisms for declaration and definition of inter-component dependencies on the programming language level that is mimicked somehow by the constructs such as Java annotations.

Although both approaches pretend to solving complex problem of *a priori* unknown architectures, in the practical approaches it seems that they are not competitive, because of completely different level of abstraction followed during design of the system. Although both agents and components may be perceived as abstract and system decomposition entities [20], in the former case the substantial feature becomes the logical acting connected with the way of realization of the requirements by the interactions focused on local goals, while in the latter case the stress is put on the implementation aspects, focused on the way of delivering to the system necessary functionalities [185]. So if the agents are analyzed from the point of view of their goals, acting strategies, required information, way of acquiring the information, or communication protocols, then this analysis is carried out in the categories of the abstract entities of the agent-based system. However when the way of implementation becomes important, in the context of conformance and completeness of technological solutions, one can consider decomposition entities that are described in the categories of required and delivered dependencies, they can be treated as components, not necessarily identified with agents. In this case one can imagine that all the agent-based implementations might share the same implementation of the communication protocol delivered to the execution environment as a component. So in this case, the granularity level of the implementation decomposition will be different to the abstraction level of the system analysis conducted in the agent-based categories, that will be realized using possibly many components [233].

Therefore it seems that the agent-based and component-based approaches are not contradictory but rather can be easily used in a complementary way that will be illustrated by the considerations of the technological nature in the following chapters.

Chapter 5
Towards the Implementation of Agent-Based Computing Systems

Development of systems based on the already presented ideas of agent-based computing requires relevant tools supporting their design and execution, adequate to the required problem scale. The preliminary analysis of the features of the available agent-based platform and simulation frameworks has been conducted. It showed clearly that available frameworks are inefficient and inflexible in the context of the considered computing applications. In other words, none of them can be used both for construction and efficient realization of the distributed computing focused on processing a large number of agents, using robust communication mechanism that are required for such class.

In order to show the efficacy of the proposed concepts, it is necessary to design and develop structures and basic services encompassed in a possibly universal framework (platform), dedicated for relevant computing and simulation class, leveraging the concept of agent, in particular focusing on agent-based, as well as classic, population-based computing models.

This chapter presents the study of possibilities of utilizing of existing frameworks for realization of agent-based computing systems in the perspective of the presented architecture. It becomes a starting point for detailed description of the requirements and technological assumptions, as well as the assessment of the possible development variants of agent-based framework and distributed computing environment.

5.1 Technological Background

Considerations presented in the previous chapter show, how many different systems are counted into the class of "agent-based" or "multi-agent" ones. As a consequence of this diversity, different technological perspectives of the particular system classes, followed by a vast differentiation on the tools required for their development. From

© Springer International Publishing AG 2017   123
A. Byrski and M. Kisiel-Dorohinicki, *Evolutionary Multi-Agent Systems*,
Studies in Computational Intelligence 680, DOI 10.1007/978-3-319-51388-1_5

this point of view, it seems that it is possible to identify of the two following application classes:

- *information* systems consisting of cognitive (deliberative) agents, often oriented on sharing services in open and heterogeneous environments, using sophisticated communication languages and complex interaction strategies, becoming the application field of so-called agent-oriented technology (cf. Sect. 4.1),
- *simulation and computing* systems, consisting of relatively high number of computationally-simple, reactive agents, acting in parallel, but rather not oriented on execution in the network environment, using simple communication strategies based on broadcasting of information in local environment determined by the system structure (cf. Sect. 4.1.2, to this category the previously described agent-based evolutionary computing can be counted.

The main criterion of this differentiation is the system structure and its development technique, and connected to it relative weight of the infrastructure. The system infrastructure is considered as all system's elements, e.g. communication support, registration and directory services, management mechanisms etc.

5.1.1 Referential Solutions

Agent-based **information** systems, though their distributed implementation are usually characterized by a flat organizational structure (see Fig. 5.1), caused by equivalent treating of all agents and a global addressing system used. It can be seen mostly in the communication structure, which range is limited only by the efficiency of the directory information passing. Such a system structure is not contrary at all to the fact of arising of virtual organizations connected with existing of certain roles assigned to the agents and realization of appropriate interaction strategies (cf. [139]).

Systems belonging to this class have usually very complex implementation, but *relative* weight of their infrastructure is small, especially when the common assumption that "everything" is an agent (see the description of FIPA standard and JADE platform in Sect. 4.1). Surely agents here have the highest development complexity and their implementation affects the way of realization of the system's goals. As an effect, the infrastructure "knows" not much about the states and the way of execution of the agents, and this knowledge is usually limited to the agent's identification and information about its services [85, 171].

In the case of agent *computing and simulation* systems, the implementation structure reflects the system organization, usually a hierarchical one (see Fig. 5.2), on one hand hiding the fact of possible system distribution, on the other hand delivering a natural way of limiting the communication range. Systems belonging to this class have usually less complex implementation, but their *relative* infrastructure weight is high. It is because of the efficiency of the realization of the system's goals when large number of agents. In order to maintain the efficiency of the system execution, the

Fig. 5.1 Structure of the information agent-based system

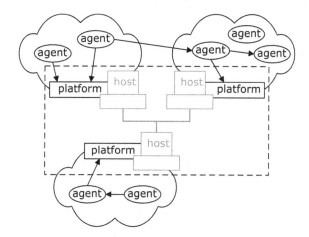

Fig. 5.2 Structure of simulation (computing) agent-based system

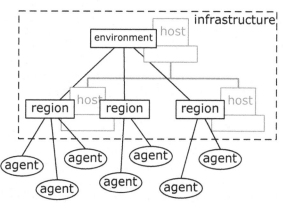

infrastructure takes over the execution of many agent-related functions, still leaving to them a certain level of autonomy, especially when considering the choosing of the action to be executed. The agents can even be perceived as having limited executive autonomy, assuming full decisive autonomy, as it is realized in e.g. Swarm or Repast (cf. Sect. 4.1.2), where *de facto* agents are managed by an event schedule. The infrastructure in this case is believed to have a significant "knowledge" about the states and individual characteristics of particular agents.

Based on the deliberations presented in Sect. 4.1, active technological support is maintained only for the first of the above-mentioned classes of agent-based systems. The platforms dedicated for open information systems are of course not suitable for realization of the computing systems. In such platforms (e.g. JADE, cf. Sect. 4.1.1) the agent is usually implemented as a thread, therefore in practical applications where hundreds or thousands individual are needed, such an approach will become inefficient. Moreover, the functionalities delivered by the platform that are relevant for open heterogeneous network frameworks (e.g. registration of agents' services, open communication languages, conformance to the standards), will become an unnecessary burden in the case of computing systems.

5.1.2   A Need for New Computing Platforms

Existing tools belonging to computing and simulation systems described in the previous subsection are usually used for development of simulation systems (cf. Sect. 4.1.2). Such frameworks allow creation of lightweight and efficient agents, that however do not have typical support for all agency features (e.g. addressing and communication). Moreover, in most cases such platforms do not support distributed computing.

It turns out, that although the ideas of EMAS naturally can be located in the simulation and computing class of the agent-systems frameworks, they in fact require tool that encompass selected solutions present in both described classes, including:

- possibility of implementation both lightweight, computationally efficient agents (similar to the ones present in Swarm), as well as heavyweight agents, focused on communication and fully autonomous (similar to the ones present in JADE),
- addressing of agents and support for asynchronous communication,
- organization of agents in local environments limiting their interaction range,
- possibility of system distribution and agent mobility.

The presented aspects of management of computations and the discussion about possibilities of using of the existing tools, encourage for searching own technological solutions appropriate for building efficient and flexible agent-oriented computing and simulation systems. The presented idea of agent-based evolutionary computing assumes hierarchical logical structure of the system, that may be treated as extension of the classic migrational model of parallel evolutionary algorithm (cf. Figs. 1.2 and 2.6). The most important elements of this structure are [53]:

- Environment: supreme element, responsible for management of the islands and bridges,
- Island: distribution element and abstraction of the local environment, particular islands are implemented as independent threads, communicating between themselves asynchronously,
- Bridge (path): directed connection between islands (a channel through which the communication is realized),
- Agent (individual): basic active element of the system implementation, located on particular island,
- Resource: passive system element representing data or functionalities used by the agents.

Focusing on these features, a number of tools were designed sharing similar organizational structure, but using different technological solutions, making possible building of distributed computing systems (both for classic and agent-based population computing models), as well as for simulation purposes (e.g. AgWorld, NEvol and Ant.NET platforms) [27, 172]. The mentioned computing tools were focused on efficient communication and interoperability, while the simulation frameworks put stress on adaptability to the problem variants and possibility of dynamical reconfiguration of the system.

The realized systems were used mostly for running a broad spectrum of computing experiments, making possible deriving different conclusions regarding both the ideas of the algorithms run, as well as the design and technological assumptions. The main disadvantage was too small flexibility and extensibility of the implemented tools. It was caused by a very strong binding to the technologies used (e.g. communication) and to assumed implementation structure, making the development and further adaptation of the complete system a very difficult task requiring reimplementation of almost the whole system. However, the experiences gathered during the development of these platforms were applied in the course of design and development of a more universal computing and simulation framework.

5.1.3 Basic Requirements and Assumptions

In order to run the considered systems supporting population-based computational intelligence techniques, it is important to consider the following aspects, that may be perceived as non-functional requirements:

1. *Efficiency.* There may be a huge number of agents and a huge number of simple messages (queries) among them.
2. *Flexibility.* The architecture is general enough, in order to become a base for developing a broad class of systems, retaining high cohesion and minimization of the coupling between its components.
3. *Extensibility.* It is necessary to easily define new system elements (e.g. agent types), preferably realized by putting up together the available fragments based on an existing configuration.
4. *Distribution.* The system works in parallel on many computing machines connected by a network (a computing cluster), however the communication among the computing nodes in independent from a certain protocol.
5. *Manageability.* Mechanisms for monitoring of certain agents, their properties and actions are needed. The system may be monitored and managed by many actors at the same time.

In addition, the following particular assumptions have been accepted:

- The concept of the system is independent from certain programming language, operating system or hardware platform.
- The new system is developed based on existing solutions provided by supplied libraries (repositories) and by implementing of well-defined interfaces.

- The system configuration is saved in a legible form and it can be easily modified. It is possible to configure in a simple way the important fragments (e.g. agents) and to integrate their definition with the existing configuration.
- It is possible to develop user implementations of communication mechanisms in a distributed environment, management components etc.

5.2 Agent Design

Following the software design perspective, an agent can be easily described using notions from object-oriented paradigm: similarly to an object, the agent has its state and can execute certain services as answer to incoming messages [298]. Thus, however the analogy is finished, as the most important difference between agents and objects is their autonomy level. Though the object has certain autonomy, in particular concerning its state (providing it has been designed according to the encapsulation rule), it is not autonomous in its behavior, contrary to the agent. The reason for this is that in the object-oriented programming, the decision whether to execute a service of an object is undertaken by another object, sending to the former one a certain message. Meanwhile in the agent-system, the decision how to react on the received message is undertaken by the agent. Agent knows its goal by-design and object solely executes its services. This is well summarized in the following phrase: "objects do it for free, agents do it for money" [154].

5.2.1 Agents in Object-Oriented Paradigm

Agent-oriented paradigm can be easily realized using the object-oriented approach (e.g. [167]), and in most cases this is the way. Of course because of the mentioned analogies, agent is usually represented by a single object, but its implementation uses many different objects, thus constituting a small subsystem (reminding a popular Façade design pattern [124]), that was presented in Fig. 5.3.

Agent's knowledge consists of elements of the state of the objects representing the agent and can be retrieved from the environment by observation (e.g. accessing the directory service), or communication with other agents (realized by the means of infrastructure of the platform, according to the Mediator design pattern [124], as it was shown in Fig. 5.4) Agent's actions are realized by the elements of the implementation of the methods of "its" objects, calling the platform infrastructure services, wherever it is necessary. Plans, strategies and goals of the agent can be expressed both in the form of the state and the structure of implementation of "its" objects. Because the basic structure of the agent is delivered by the platform, its specializations devoted to realization of particular applications should be realized according to well-known design principle OCP (Open-Closed Principle) [202], using

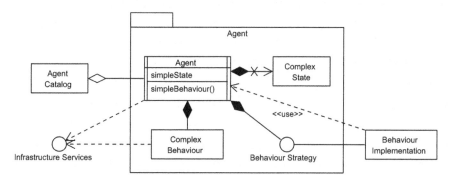

Fig. 5.3 Schematic illustration of agent implementation structure

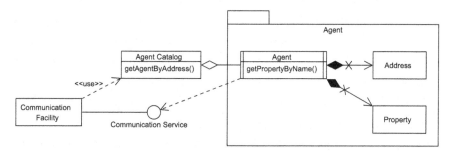

Fig. 5.4 Implementation of communication between agents using the underlying platform infrastructure

design patterns such as Template Method or Strategy [124], as it was shown in the case of `Strategic Behaviour` in Fig. 5.3.

Because of the memory management, coupling between objects of the subsystems of different agents should be minimized. In this case, using a dedicated addressing mechanism, that is independent of identities of agents' objects or even code instrumentation of the agent's class, allowing for accessing particular objects by using identifiers, can be very helpful.

The simplest instrumentation can consist in expressing the features of the agent by an uniform interface, making possible accessing its properties by names,[1] particularly pointing out the property containing an unique address of the agent (see Fig. 5.4).

One must however remember, that one of consequences of using such mechanism, based on constant applying of reflection, can lead to serious reduction of the efficiency of the system. To sum up, though instrumentation used for sake of management of the computing makes sense, in the case of implementation of the agent system logic, a compromise should be sought between flexibility of the architecture and efficiency.

---

[1] Such solution is compatible with techniques supporting management of applications, e.g. JMX in Java [184].

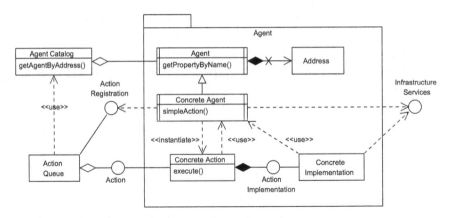

Fig. 5.5 Implementation of the agent's activities using the action mechanism

In particular cases code profiling may be necessary to check the overall efficiency of the application.

The agent can decide on its own about execution of an action in its environment. Note that though similar processing cannot be realized in the classic object, the agent may be in such case perceived as an "active object" [57]. In a multi-threaded architecture, agent can be implemented as object-thread, making necessary taking into consideration concurrent features of such system. Because in majority of models and practical realization of the agent systems the agent's actions are modeled in discrete time, it is much simpler and more effective to employ in such case discrete simulation (according to the event or phase model) [243]. In order to utilize contemporary multi-core processor architectures, it seems to make sense to hybridize concurrent, thread-based realization and discrete simulation. Such an approach seems to be natural in the case of hierarchical structure of the system (lower hierarchy levels utilize threads, and inside the threads discrete simulation of agents is realized). Because the ultimate outcome of the execution of an action depends on the context (state of the environment in the moment of its execution), in order to assure independence of agent and environment, one should utilize the design pattern Command [124]. If at the same time, specialization of the execution of the action is necessary, one can also use the delegation techniques, similarly to Strategy design pattern, however a care should be taken, because potential high coupling between the action code and agent can occur (see Fig. 5.5).

The above assumptions are well suited for the basic architecture of the agent-based computing system. This is connected to the postulate of specialization of agent's implementation by the means of delegation rather than by inheritance, thus realized according to the Strategy design pattern (see Fig. 5.6). It makes possible to follow the Liskov Substitution Principle (LSP) [190]. The hierarchy of agent systems, may be implemented as an aggregate agent that manages the lower-level subsystem and realizes the commitments in its own environment. It can be thus perceived "from outside" (e.g. from the perspective of the system manager) as realization of the Composite

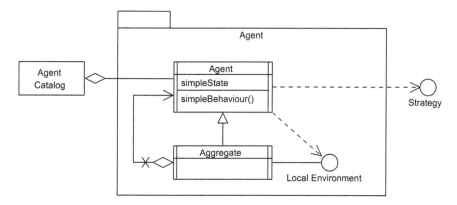

Fig. 5.6 Idea of implementation of the hierarchy of agents

design pattern. The agent however should make available only absolutely necessary interfaces in order to avoid too strong coupling (following Interface Segregation Principle (ISP) [202]). One of such important interfaces is `Local Environment` in Fig. 5.6. Unfortunately this contradicts the assumptions of high cohesion, so one can consider decomposition of such class into strongly coupled services oriented on particular aspects of management and system logic.

One must remember that the components described above, though easy to realize, are not an integral part of classic object oriented model and require supplying of additional elements of the implementation framework, that can be treated as infrastructure of so-called agent platform (cf. Sect. 5.1) represented on the above-shown exemplary schemes by the such elements as: `Agent Catalog`, `Infrastructure Services`, `Communication Facility`, `Action Queue`, etc.

5.2.2 Component Structure of the Agent Implementation

The deliberations presented in previous chapter were summarized by the statement, that similarly to agent paradigm, component techniques try to answer the problems of building complex systems having at the time of development unknown architecture. It can be said, that from a certain point of view, both agents and components deliver the support for delegation of responsibilities, but in the different level of abstraction. It can be said, both agents and components support the delegation of responsibilities, but on different levels of abstraction. In the case of agents, realization of complex interaction procedures requires using of common communication protocols, and usually also directory services, allowing for identification and localization of agents (interoperability). In the case of components, the key aspect of code reusability requires solely the possibility of calling the known services delivered by other components. As a consequence, it seems that agent as a decomposition unit

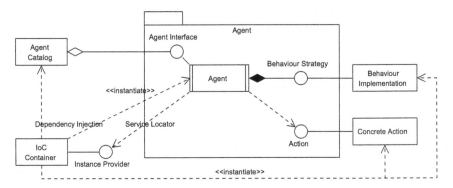

Fig. 5.7 Example of bindings between configured objects using dependency injection and dynamic location

and at the same time being part of the system will have much more complex implementation, because of necessity of realization the requirements of the platform, than component, which in extremal case can be implemented as a single class [20].

Agent based system can be assembled using IoC container based on configuration describing among other bindings between properly initiated objects of the selected classes, describing particular elements (agents, algorithms, actions). Configuration of the system can be expressed based on a dedicated architecture description language or using an application programming interface. From the configuration point of view, each single element of the system implementation can be treated as component, and its specification as a complete description containing identifier and definitions of required and provided dependencies. Based on the component specification, it is possible to verify the correctness of the system configuration during initialization of the component framework and subsequent assembling of the complete system using particular elements (see Fig. 5.7).

Following this approach the framework guarantees that after creating of the component instance, it has access to all required dependencies, fulfilling the assumptions of functional integrity of the system.

Dependencies of the component may be defined by the type system of the programming language (e.g. interface names) or component identifiers. Using of type system makes possible utilizing of the embedded generalization relation, and auto-wiring of the components, described in Sect. 4.2.2. On the other hand, the identifiers of the components make possible using of one class as a base for different definition of the required dependencies. Thus from the configuration point of view this approach makes possible delivering different components referring to the same implementation (this mechanism is similar to parametric or generic types available in many programming languages). The specification of the component can become an integral part of the code of the class, if expressed using appropriate naming convention (e.g. in so-called setter injection) or using metadata (e.g. annotations in Java). It may be also expressed as an additional specification (so called manifest), that can be written in a form readable for the user, using so-called domain-specific language [119]. Of

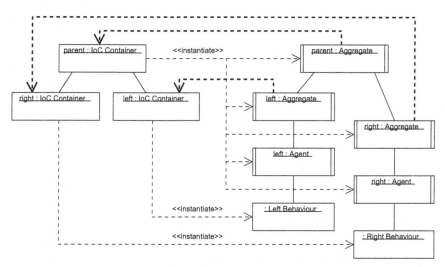

Fig. 5.8 Hierarchy of IoC containers and agent systems presenting the so-called robot legs problem

course only in such situation it is possible to declare many components referring to the same class.

Considering the hierarchy of agents it can be assumed, that implementations of various agents in hierarchy can differ, or can be composed of the same components. In order to solve this problem one can assume existing IoC containers for each agent subsystem, that also can be organized in hierarchy, minding the dependency relation, allowing to access the components registered in higher-level IoC containers (this is shown in Fig. 5.8, and sometimes is described as "robot legs problem").

In this situation the configuration of the system should also describe the dependencies between the agents, making possible building hierarchical structure of agents and adequate structure of IoC containers [234].

There is a certain flaw in the described component model, namely, in many cases there will be a strong coupling in semantically connected groups of elements (algorithms and actions) realizing complementary tasks (e.g. components of different operators for the same representation of the solution used in search heuristics, cf. Sect. 1.1). Because of this one can consider two-level organizational structure of the implementation where such strongly coupled groups of elements become an upper-level composition entities. Because of terminological problems discussed in Sect. 4.2, they will be simply named "modules". It is to note, that these modules may be also used as a basis for the organization of the process of building of the parts of the system using tools such as Maven.

If assumed, that the configuration will also contain the parameters of the objects instantiation, it will become a complete "recipe" of a concrete running of a system, that will help in management of the deployment process. In many situations it might be necessary to make advantage of dynamic object binding, therefore in order to use the information read by IoC container, it should be equipped with on-demand instantiation mechanism—realizing in such way Service Locator [6] design pattern (also shown in Fig. 5.7). In such case, to specify a concrete service one can use both component identifier, as it's type (name of an interface or of a class). As the component instance is delivered also by IoC container, that is responsible for initialization and verification of the correctness of particular instance, it will be also equipped with all required dependencies.

5.3 Computing Environment Architecture

The above presented deliberations encompassed development aspects connected with efficient, flexible and extensible implementation of computing agent-system architecture. However certain aspects still need to be discussed, namely the ones concerning distribution and management of computing, adequate to possible perspectives of its utilization. It means practically that besides elements closely connected with logic (architecture and behavior) of the desired system class, the following functions must also be considered:

- delivering implementation and configuration components,
- managing computing and communication in the distributed environment,
- computing monitoring.

These functions may be encapsulated in modules deployed in possibly distributed nodes of the computing environment.

5.3.1 Computing Node

According to the Separation of Concerns rule, it can be assumed, that the above-mentioned aspects can be separated on the implementation level from the logic, possibly placing them in separate modules, that might cooperate with themselves using by mutual sharing of services. In Fig. 5.9 an exemplary implementation structure of a single instance of computing framework was shown (from now on named computational node, for the sake of simplicity, or simply a node) along with pointing out the dependencies between particular modules. In accordance with flexibility and extensibility requirement, the proposed modularization suggests the possibility of applying component-based techniques in a similar way as on the configuration level, making possible cooperation of different implementations, if only the dependencies are structurally the same. Of course the configuration and managing of cycle of mutually-dependent services is a consequence of such approach, that must be taken into consideration when choosing a particular technology.

The most natural solution in the case of designing the computing node seems to be using the same technology, as for the computing itself (agent hierarchy), namely IoC

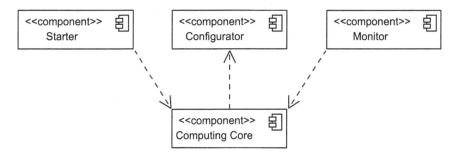

Fig. 5.9 Exemplary implementation structure of a computing node

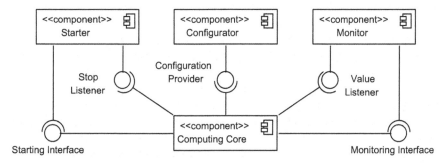

Fig. 5.10 Exemplary implementation structure of a computing node with explicitly shown interfaces

container. It requires definition of a class representing each module for the dependency configuration purposes (the already mentioned Façade design pattern), with provided and required interfaces (see Fig. 5.10). Because the complexity level of implementation of particular components of the computing node is much more complex, resulting in limitations of this approach that can become seriously problematic in integration of different technologies used by the node components. In such case one may, and even should consider using other solutions, more oriented on modularity. As an alternative or complement of the approach based on IoC, one can use modular techniques, such as the ones described by OSGi standard (cf. Sect. 4.2.3).

5.3.2 Distributed Computing

The communication support is the most important requirement of the computing realized in distributed environment. In general, one can assume, that each machine realizes part of computing as one process of the operating system, that can be identified with the above-described computing node. In this situation, the most important thing is identification of different computing nodes in the frames of one computing

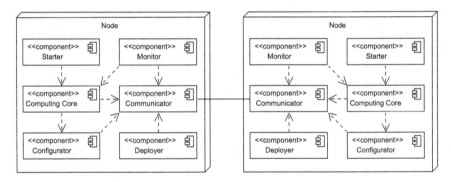

Fig. 5.11 Exemplary structure of dependencies of the node communication services

system. Addressing of distributed processes can be natural for the applied technology (e.g. MPI), but many solutions is based rather on the host addresses or identifiers of services or data structures, as it is realized in the case of brokers or directory services, e.g. JNDI. However, most of currently used technologies assumes addressing that is transparent for the user. Assuming that there is no need to follow one concrete protocol nor communication technology, it seems to be necessary to assume an independent method for addressing the nodes and equipping the communication mechanism in capability of mapping the "internal" addresses of the computing environment to the node identification method used in concrete technology.

After the above assumption of component-oriented realization of the node structure, the communication in the distributed environment can be the subsequent module delivering services appropriate for the selected service communication paradigm. Following the separation of concerns rule, it seems to be correct to encapsulate the issues concerning the communication behind a well-defined interface—in this way the implementation of the remaining modules will not depend on a plethora of technological solutions that are available in the area of distributed communication. It was shown schematically in Fig. 5.11 using the same modules as in the previous subsection, assuming that the `Configurator` and `Monitor` are mutually aware of many distributed instances of the computing, and `Deployer` assembles the computation itself.

The main client of the communication services will be of course the component realizing the computation logic, thus because of the characteristics of the communication between the agents, it seems that as the reference point for the construction of communication interface, one should assume the message passing model. Of course the implementation of the transport layer that is internal for communication services can efficiently use other communication paradigms, e.g. in the context of broadcast messages requirement, a very efficient implementation can be achieved using publish–subscribe model (e.g. ZeroMQ), or emulation of shared memory (e.g. JGroups), while management communicates can be easily implemented in the remote calling techniques (e.g. RMI or ICE).

Such an approach as the one described above may assume peer to peer cooperation of services running on different nodes, or any model of logical cooperation of services. From the management of computing point of view, e.g. delivering configuration, startup or monitoring, it can be master–slave model (or hierarchical one), where the master node will manage the whole computing (e.g. using the central repository or by the GUI). Such solution is of course very simple to realize from the implementation point of view, but at the same time bears a significant flaw, namely the master becomes a single point of failure. Other, more typical way of realization of master–slave model is delegating of time-consuming, independent operations realized by the agents (e.g. computing fitness with high complexity), to remote computing nodes. A very interesting application of such computing model is leverage of the computing power of voluntarily shared computer systems: volunteer computing, that was shown in the case of applications run on WWW browsers [42].

Following the above assumptions, other aspects of system distribution, such as:

- assembling and reconfiguration (adding, removing nodes) of the distributed computing environment,
- running, monitoring, acquiring of the computing results,
- virtual communication topologies,
- load balancing.

can be realized by the dedicated modules deployed on the computing nodes and cooperating with themselves. Such an architecture is an extension of the applying the separation of concerns rule to the distributed environment, at the same time contradicting somehow Service-Oriented Architecture paradigm (SOA is not a very good solution for computing system because of efficiency-related issues).

A very important issue to be discussed in the perspective of the presented deliberations are possibilities of distribution of the proposed computing architecture (agent hierarchy). Following the prototype solutions (cf. Sect. 5.1.2), it seems to be the most natural to map the system agent residing on the first hierarchy level on the distributed node structure and treating each of them as a root of a local agent hierarchy (see Fig. 5.12).

This approach causes that the main hierarchy element delivering the environment to the agents must be "virtualized", i.e. its functionalities must be taken over by distributed instances of computing implementation infrastructure on the node level. It makes a very useful tool for load balancing, as whole (sub)trees representing parts of computing can be assigned to nodes in many ways. An alternative possibility is mapping of random (sub)tree in the hierarchy (or even a single agent) to other node (so called bilocation of agents, as it is presented in Fig. 5.13) that could give particularly interesting results in the context of the above-described master–slave model in volunteer computing environment [42].

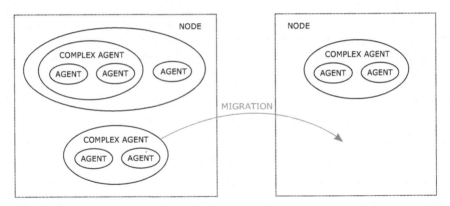

Fig. 5.12 Flat distribution structure making possible realization of load balancing based on migration of the agents

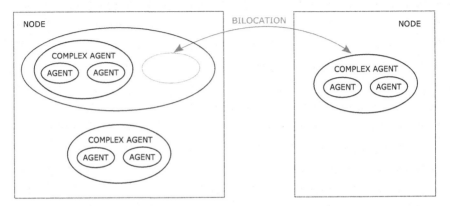

Fig. 5.13 Unbalanced distribution structure allowing for delegation of agents from random hierarchy levels

5.4 Summary

Though the above-presented deliberations concentrated on application of the agent oriented approach on the computing level, the solutions described in this section show that it is possible to encompass with a consisted technological model both application of agents to managing of computing (as it was shown e.g. in [286]), as in the computation structure. The analysis of the contemporary tools (in particular agent-oriented platform) in the context of the system architecture presented in the previous chapter, leads to assumption that can be later mapped to object structures of the programming languages with natural support of the component techniques and communication technologies. In the next chapter, analysis of possibilities of realization of the presented ideas was given, based on the structure of the actual computing platform AgE.

Chapter 6
AgE Computing Environment

AgE environment[1] has been developed as an open-source project at the Intelligent Information Systems Group of AGH-UST. *AgE* provides a platform for the development and execution of distributed agent-based applications—mainly simulation and computational systems. The most advanced version of the environment was developed using Java (jAgE), but there were also many additional prototype environments realized e.g. based on .NET, Python, Scala, Erlang and Haskell.

Figure 6.1 presents an overview of a system based on *AgE* platform. A user constructs the system by providing an input configuration in XML format. The configuration specifies the simulation structure and problem-dependent parameters. After the system start-up, the environment (agents and required resources) are instantiated, configured and distributed amongst available nodes where they start performing their tasks. Additional services such as name, monitoring, communication and topology service are also run on these nodes to manage and monitor the computation. The output of the simulation is problem-dependent and may be visualized at run-time by dedicated graphical tools and interpreted by the user.

6.1 Agent Platform

In the core of AgE there is an execution environment which allows to design the computation model, i.e. the agents and their interactions. This part of computing environment is functionally similar to agent platform, yet needs to be supported by other services to be properly configured, initialized and deployed in specific (e.g. network) settings.

The basic element of the platform abstraction is an agent (required interface IAgent). Each agent has an unique (in the whole system) address used as an

---

[1]http://age.iisg.agh.edu.pl.

© Springer International Publishing AG 2017

A. Byrski and M. Kisiel-Dorohinicki, *Evolutionary Multi-Agent Systems*,
Studies in Computational Intelligence 680, DOI 10.1007/978-3-319-51388-1_6

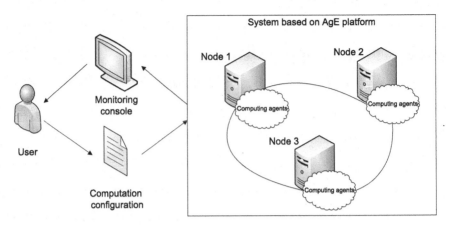

Fig. 6.1 AgE system overview

identifier in one execution of the environment. The agents can be divided into two types, a simple ones and aggregates (`IAggregate`) that can "contain" other agents cooperating among themselves, in fact creating another, embedded agent-system. Such agents, besides their own activities, manage also the work of the "subordinated" agents, i.e. mediating into communications among them, managing their life cycle, action execution etc. The agents are thus organized into a hierarchy (see Fig. 6.2), similar to the one present in the Swarm platform (see Sect. 4.1.2).

Each agents acts in an environment that may be treated as a kind of its context, delivered by the superordinate agent, which provides the subordinate agent also with necessary information, resources and services. The main agent (residing on the top of the hierarchy) is an exception here, representing the local computing environment

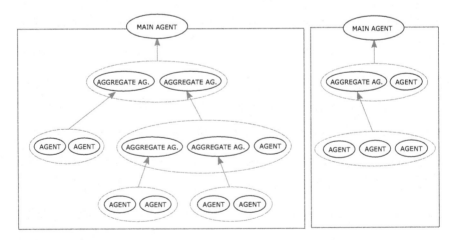

Fig. 6.2 Agent tree structure

(interface IWorkplace) and which is provided by the information and services by the infrastructure of the computing node. The subordinate agent does not have a direct access to the superordinate one, but can communicate with it using a predefined interface: ILocalEnvironment, that delivers the following functionalities:

- order of execution of certain action, e.g. sending a message to another agent,
- querying the properties of agents (observation), that may consider the following two areas:

  – agents residing on the same hierarchy level (having the same superordinate agent),
  – agents residing on the level of the superordinate agent,

  while it is assumed, that on the higher level, only selected information are available,
- registering a new agent in the system (possibility of performing of such action is verified with all the agents in the hierarchy—when the request reaches the main agent it may allow or block the creation or migration on the lower levels, allowing for management of the system efficiency).

6.1.1 Execution of Agents

It is assumed that all agents at the same level of the structure are being executed in parallel. The platform introduces two types of agents: thread-based and simple. The former are realized as separate threads so that the parallel processing is managed by Java Virtual Machine (similarly to JADE platform). Such agents can communicate and interact with neighbors via asynchronous messages. However, a large number of such agents would significantly decrease the performance of a simulation because of frequent context switching and raises synchronization problems. Therefore, following the concept of *phase simulation*, the notion of simple agents is introduced. The execution of simple agents is based on *steppable* processing which is to simulate pseudo-parallel execution of agents' tasks. Two phases are distinguished:

- Execution of tasks related to computation semantics in the step() method. In case of an aggregate agent all it's children perform their steps sequentially. While doing so, they can register various events, which may indicate actions to perform or communication messages, in the parent aggregate.
- Processing of events registered in an event queue. Since events may be registered only in agents that possess children, this phase concerns only aggregate agents.

The described idea of agents processing ensures that during execution of computational tasks of agents co-existing at the same level in the structure (agents with the same parent), the hierarchy remains unmodified, thus the tasks may be carried out in any order. From these agents perspective, they are processed in parallel. All changes to the agent structure are made by aggregates during processing of the events that indicate actions such as addition of a new agent, migration of an agent, killing an already existing agent, etc. They are visible for agents while performing the next step.

The environment of simple agents determines the types of actions which may be ordered by child agents. It also provides concrete implementations of these actions and thereby supplies and influences agents' execution. Thus actions realize the agent principle of goal level communication [21], because agent only lets the environment know what it expects to be done but it does not know how it will be done.

Simple agents request their parent aggregates to execute actions during an execution of a step of processing. Then, all of actions are executed sequentially (in order of their registration) by the aggregate after all children agents finished their operations.

Because some of the actions can significantly change the environment (for example removal or migration of an agent) so that the other actions would become invalid, the following phases have been introduced:

1. initialization (init), when target addresses are verified,
2. execution (main), when the real action is executed,
3. finalization (finish), for performing activities that could not be executed during the main phase (e.g. removal of an agent when other agents could refer to it).

All changes of agents structure that can influence execution of other registered actions are performed in the finalization phase. As a result, performing an action in the execution phase is safe. In the initialization phase actions can perform some preparation activities that are required by other actions.

Two types of actions exist:

- Simple actions that can define only one task to be performed on only one agent.
- Complex actions—they are containers for other actions and can hold a tree-like structure. Actions wrapped by them are executed in a well-defined order and allows to create more complicated scenarios like an exchange of resources, when the separate component actions are required for getting a resource from one agent and for delivering it to another.

Most simple aggregate actions are defined as methods in a class of an aggregate agent and the default aggregate implementation provides some actions out-of-the-box:

- adding of a new agent,
- moving an agent to another aggregate,
- death of an agent,
- cloning of an agent.

Moreover, users can extend the platform with any actions they need. These actions can be created as strategies bound to the aggregate using the configuration of the platform. They allow to extend functionality of the platform in an easy way but have a downside of not having the possibility to refer to private members of the aggregate. Decision of how to execute such actions is made by the parent agent who resolves proper action implementation according to *Service Locator* design pattern [6].

6.1.2 Life-Cycle Management

The lifecycle of an agent consists of the following phases:

1. Construction — when a constructor of agent class is called.
2. Initialization of the object dependencies and properties — when the init() method is called; at this point the agent has all its dependencies injected by the component framework based on dependency injection pattern mechanism. Also its properties are initialized using the component framework or by agent itself. For example at this stage, an agent generates an address.
3. Initialization of the environment — the moment when the parent of the agent calls the setAgentEnvironment() method. At this point the agent can use mechanisms that requires the existence of the local environment i.e. actions, queries, messaging.
4. Finalization of the agent — the finish() method. The agent should finish its operation at this point.

Threaded agents additionally provide the run() method, called by the Java Virtual Machine after their dedicated thread was started. At this moment they can start the main loop of their execution.

The full lifecycle of the simple agents is shown in Fig. 6.3. Simple agents need to provide an implementation of the step. It is done in the step() method. This operation is called in an arbitrary order by the parent aggregate on every agent it contains. The actual execution from the point of view of the whole tree of agents is performed in the postorder way: firstly the aggregate lets children to carry out their tasks and only after they finished them it executes its own tasks.

During the execution of the step, the simple agent usually needs to perform following actions:

- receive and send messages,
- execute queries,
- execute a part of the computation,
- order actions for the parent.

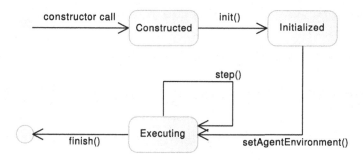

Fig. 6.3 A life-cycle of the simple agent shown as a state diagram

Fig. 6.4 Components of an
agent address

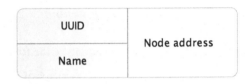

After iterating over all children, the aggregate needs to process the event queue.
These events are usually actions requested by the children.

6.1.3 Communication Facilities

The platform allows for all agents to have a unique addresses, which allow for
their identification and supports inter-agent communication. The particular property
of being globally unique is guaranteed by a structure of the address. As shown in
Fig. 6.4, the agent address comprises of three components: an UUID,[2] a node address,
and a name. Two former parts identify an agent in the platform instance and the last
one is a replacement for an UUID provided for the user convenience (for usage in a
user interface or logs).

An address is usually obtained by an agent during the initialization of the component dependencies (see page xxx for the explanation of the agent life-cycle). It
is done by requesting a new address object from the `AgentAddressProvider`
instance that is a local component of a node.

Communication via message passing

Agents located within a single aggregate can communicate with each other via simple
messages.

Interfaces used in messages are shown in Fig. 6.5. A message defined by the
`IMessage` interface consists of a header and payload. The header, as defined by
the `IHeader` interface must specify a sender of the message (usually the agent
that created the message) and its receivers. The payload is simply a data of any
(serializable) type that is to be transported.

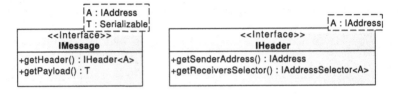

Fig. 6.5 Overview of interfaces used in messaging

---

[2]Universally Unique Identifier.

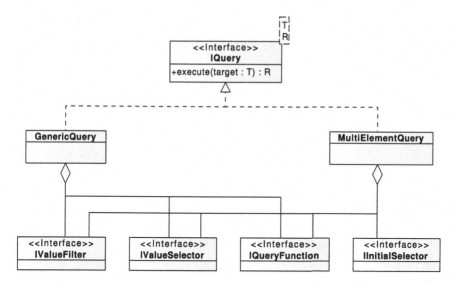

Fig. 6.6 Overview of queries base classes and interfaces

Receivers are defined using selectors. They offer a possibility to define receivers with the *unicast, broadcast, multicast* or *anycast* semantics.

In the case of simple agents, sending and delivery of messages is performed by an aggregate agent. The sender adds a message event to its parent queue. The parent handles it by locating all receivers and calling a message handler on each of them. These messages are placed on a queue and can be received by the agent during its next step.

Thread-based agents use a similar queue of messages but are not restricted by the execution semantics and can inspect it at any point of time.

Query mechanism

Queries offer a possibility to gather, analyze and process data located both in local and remote (from the point of view of the query executor) components.

The diagram in the Fig. 6.6 shows base classes and interfaces of the query mechanism along with their interconnections. The central point of this mechanism is the IQuery interface. It provides only one method: execute(). A query, as defined by this interface, is an action performed on a target object that leads to creation of query results. Specific implementations define a relationship between the target and results.

On the top of this interface and definition, a simple, declarative, yet extensible query language is built. Queries are implemented as (GenericQuery and MultiElementQuery classes in the diagram 6.6. It allows the user to perform tasks like: computation of the average value of some chosen properties from the agents in the environment, select and inspect arbitrary objects in collections and much more.

The following operations are defined:

- Preselection of values from the collection. It is only available if the query is performed over an iterable type instance. Its task is to select some (e.g. first ten or random) of objects without usage of the object-related information.
- Filtering by a value. This is an operation similar to WHERE clause in SQL.
- Selection of values. It can select specific fields from objects and it shows some similarities to the SELECT operation from SQL. If this operation is not defined then whole objects are selected.
- Functions working on an entire result set. They can remove, add or modify elements.

Operators are defined as implementation of specific interfaces (one for every operation, as shown in Fig. 6.6). They are presented to the user as static methods (e.g. `lessThan()`, `pattern()` etc.).

A query is built by specifying following properties:

- A type of the target object (the object passed as an argument to the execute method).
- A type of results.
- In the case of collections — a type of elements in a collection.

Such an exhaustive specification is required because queries rely on these pieces of information to control correctness of operators used by the user (with the usage of Java generics). Moreover, queries in AgE are built without the knowledge of the target object (it is in opposition to many similar mechanisms like LINQ[3]).

After that, an operation of the query is specified using aforementioned operations. The execution of the query is carried out by calling the `execute()` method.

The following Java code shows a simple example of how a query can be created and executed. In this case a collection of strings is queried.

```
CollectionQuery<String, String> q =
    new CollectionQuery<String, String>(String.class);
q.from(first(10))
    .matching(anyOf(
        pattern("li.*"),
        pattern("lorem[Ii]psum")));
Collection<String> results = q.execute(someList);
```

It can be noticed that queries definition uses the *fluent interface* pattern with specific operations being composed from static methods.

This approach of declaring a query without the knowledge of the target is additionally useful because it allows to execute a single query many times (possibly with caching the results or operations) or to delegate queries to be executed in another location. The query delegation is actually often used within the platform during the

---

[3]http://msdn.microsoft.com/en-us/netframework/aa904594.aspx.

operation of querying an environment of a parent of an agent. This mechanism is essential for performing the migration of agents.

The other side of the queries mechanism is the extensibility offered to the user on many levels. It is possible to create completely specialized queries (by implementing the IQuery interface), extending the described declarative mechanism or even define in-line operators when creating a query. This elasticity of queries was very important because of performance requirements resulting from some applications of the platform. An approach was adopted, that the user is able to provide much faster solutions for his specific problems.

In some cases it is also useful to let know a queried object about a query being executed on it. For this reason the interface named IQueryAware was created. By implementing it any object can communicate to the query that it wants to be notified about some events related to the execution. Currently, two events are supported: initialization and finalization of the query.

The last part of the queries mechanism is caching. The platform offers a possibility to build a cache of query results. Its expiration time is based on the workplace step counter. This cache works as a wrapper to the query (and as such is an implementation of the IQuery interface). During the execution it checks whether stored results expired and possibly executes a real query replacing old results.

6.2 Component Framework

The platform provides dedicated component framework, which is built on the top of an IoC container. It utilizes PicoContainer framework[4]—a popular open-source Java implementation of IoC container that can be easily extended and customized. IoC container based on the acquired system configuration creates the particular instances of the components, initializes the dependencies between the components and assigns the initial values to their properties. The environment makes also possible acquiring component instances on-demand, realizing the *Service Locator* design pattern. In such case the component instance is also supplied by the IoC container, that becomes responsible for its initialization and verification of its correctness.

Both agents and strategies are provided to the platform as components. Their implementation classes can define named dependencies to other components (i.e. other agents, services or any other dependent classes) and simple properties that hold for example problem-dependent values. The dependencies definition for a component type, together with class's public methods (treated as component's operations) may be perceived as requirements closely related to component contracts (as proposed by Szyperski [276]).

---

[4]http://www.picocontainer.org.

6.2.1 Property Mechanism

Properties are the mechanism of introspection allowing for monitoring and managing of the running system. Monitoring of the agent properties is realized according to the *Observer* design pattern [124], i.e. in the case of the value change, an appropriate notification is sent to the objects interested. In a similar way the agent behaviors can be monitored, such as creation of a new agent, death of an agent, cloning, migration, message sending or querying.

The properties are represented by a parametric class `Property` and are implemented as pairs (name, value) with assigned information about the value type and possibilities of changing and monitoring of the value (see Fig. 6.7). Such a description encompasses an appropriate object of the class `MetaProperty` that may be shared by many properties. The properties are shared by the component using a well-defined interface (`IPropertyContainer`), and are stored in so-called property containers (class `PropertyContainer`). In particular, the implementation of the `ClassPropertyContainer` becomes here important, that makes shares the attributes of the object of the specialization class as properties, using the reflection mechanism. It is assumed, that the agent classes are property containers, usually implemented by inheriting from `ClassPropertyContainer`. Property sets may be defined using dedicated definition managers (class `MetaPropertyManager`), while the properties of one container must be registered in one manager object.

Component properties are defined using dedicated annotations that can describe fields of methods. In order to annotate the field as a property, one should used the `@PropertyField` annotation, that defines the name of the property (attribute

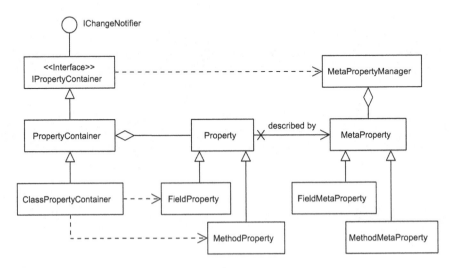

Fig. 6.7 Zarys struktury implementacji wasnoci

`propertyName`) and makes possible pointing out, if the property is monitorable (attribute `isMonitorable`), as it is in the following example:

@PropertyField(propertyName = ''genOp'',isMonitorable = true)
private IGeneticOperator genOp;

A second way of property definition is annotating the method returning its value i.e. *getter* with annotation `@PropertyGetter` and the one modifying its value, i.e. *setter* with an annotation `@PropertySetter` (the methods can be named in arbitrary way). The first annotation, besides the name of the property, has also the argument `isMonitorable`, that contains the information if the property can be monitored, as it can be seen on the following example:

@PropertyGetter(propertyName=''popSize'',isMonitorable=true)
 public int getPopSize() {
     // ...
}
@PropertySetter(propertyName=''popSize'')
public void setPopSize(int value) {
     // ...
}

If the method modifying the value is not annotated, the property is read-only.

6.2.2 Definitions of Components and Their Dependencies

Components are defined as classes that have additional information about dependencies and properties initialized by the configuration mechanism. This information is interpreted by the IoC container, that selects and injects the dependencies during the initialization of the component and assigns the values to the properties. Similarly to the typical solutions, information about dependencies are defined according to the injection method used—it is possible to do constructor injection or annotated field or method injection (cf. Sect. 4.2).

Constructors that are used for dependency injection must be public and annotated with `@Inject`, e.g.:

public class GeneticAgent {

    @Inject
    public GeneticAgent(Solution[] population) {
        // ...
    }
}

During the instantiation of the component, one available constructor is selected, which argument list fits the parameters defined in the configuration model (relevant object of `ComponentDefinition`). All the dependencies defined by the constructor are treated as required ones, so they must be initialized in order to create the object properly.

In the case of injecting based on annotated fields and methods, the initialization is realized by assigning the value of the field or calling the "set" method. Similarly as in the case of constructors, fields and methods that must be used in the process of objects initialization must be annotated with `@Inject`. By default, all the dependencies annotated in such way are optional, i.e. their initialization is not necessary for proper creation of the component instance. By using of an additional annotation `@Require` it is possible to define the desired dependency. An example presenting the dependency definition methods (required and optional) using the annotated fields or methods is as follows:

```
public class GeneticAgent {

    @Inject
    @Require
    private IMutation mutOp;

    @Inject
    private IReproduction reprodOp;

    private Solution[] population;

    @Inject
    public void setPopulation(Solution[] population) {
        this.population = population;
    }

    @Require
    public Solution[] getPopulation() {
        return this.population;
    }

}
```

6.2.3   Configuration of the System

The process of assembling a system is divided into two main phases. In the first one, the input configuration is read from XML file with well-defined structure[5] and further transformed into object configuration model, structure of which is shown

---

[5]http://agh.iisg.agh.edu.pl/age-2.3.xsd.

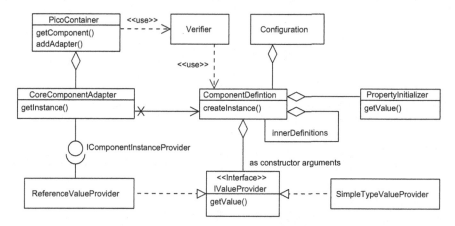

Fig. 6.8 Object configuration model

in Fig. 6.8. A `ComponentDefinition` instance describes a single component and contains data such as it's name, type (which is the name of a class) and scope, which is to determine if a new component will be created for each request (*prototype* scope) or only once during the first request (*singleton* scope). The definition also specifies the constructor arguments, which are implemented as `IValueProvider` objects and used during constructor-based injection [119], as well as property initializers, responsible for initializing component properties with reference or simple values. Moreover, the definition contains `createInstance` method which creates a new instance of a described component with initialized dependencies (this process is described below). Component definitions may form hierarchical structures (via `innerDefinitions`). If a definition is a child of another one, it is said to "exist in the context of the outer definition" and is visible only for it's parent and other parent's children (siblings). Validation of the model, performed during processing of the input configuration, allows for detecting errors such as unresolved dependencies, non-existent components or incorrect property definitions.

In the next phase of system assembly process, a hierarchy of IoC containers is built according to a structure of component definitions. For each definition a dedicated adapter (`CoreComponentAdapter`) is created and registered in a container as shown in Fig. 6.9. Moreover, the adapter implements the interface, which defines methods for retrieving instances of components by name or type — `IComponentInstanceProvider`. In the next step, verification of the configuration correctness follows, considering available component classes. The verification is conducted using the object of the `Verifier` class according to the *Visitor* design patter [124]. This object "visits" all the component definitions registered in the IoC containers IoC (Fig. 6.8). For each definition the availability of the component class and the correctness of the dependencies and simple properties are checked. In the last step, the platform infrastructure queries the base container for the list of objects

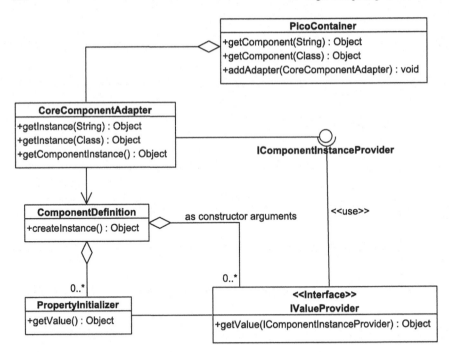

Fig. 6.9 Dependency injection in AgE platform

(agents) of the first level. During this call, the whole agent tree is initialized, along with necessary dependencies.

Default implementation of an adapter (`InstanceComponentAdapter`) instantiates during each request a fully initialized object representing the requested component (so-called *prototype scope* – see Sect. 4.2.2). It means that the adapter is also responsible for injection of all dependencies of the created component. For setting the dependencies the adapter utilizes IoC container that is delivered by a parameter of the `getComponentInstance` method. Solving the dependencies is conducted in analogically to getting the component instance, with an exception, that this time the adapter becomes the "client" of the container. The framework delivers additionally the adapter `CachingComponentAdapter`, that creates a new component instance only during the first call and later returns the previously created object, thus realizing so-called *singleton scope* in one container instance.

When a request for a component instance is directed to the container, it locates appropriate component adapter (using given name or type) and delegates the request further, to it. The adapter calls the associated component definition's `createInstance` method, which is responsible for creating a component instance. While instantiating a component the component adapter retrieves instances of dependent components from associated IoC container (or its parent), and the loop whole process starts again. In the case of simple types, a value is kept directly in a value provider object and is returned on a request. The whole process is repeated until all dependencies are resolved and then the fully-initialized component instance is returned to the client.

The presented mechanism gives a possibility to build various structures of agents with their dependencies and initial properties values based on the input configuration.

6.3 Node Architecture

The simulation is executed in a distributed environment comprised of nodes connected via communication services. Each node is a separated process being executed on a physical machine. Nodes are responsible for setting-up and managing an execution environment for agents performing a simulation, as well as assuring communication and collaboration in distributed environment.

The main part of the node is a service bus that realize *Service Locator* design pattern. The bus is realized by *AgE component framework*, which utilizes IoC container to create and initialize an object that is a run-time instance of a service. Services are being registered in the container by the node boot-strapper or other services, based on component definitions, created using API or read from XML configuration file. A reference to a service instance can be acquired by service name or type via IComponentInstanceProvider interface.

6.3.1 Node Services

The node distinguishes stateless and stateful services. The former offer functionality dependent only on parameters given in method call, therefore they does not hold any state and are always thread-safe. They are created on demand (at the first reference) and than their instances are cached in the container.

On the other hand, an instance of stateful service can hold data that influences its behavior. Such services implement IStatefulComponent interface, which introduces init and finish methods, called by the service bus while creating and destroying a service instance. Instantiation and initialization is performed during node start-up. Stateful services can be also realized as threads, that are started in init method and finished asynchronously while destroying the service.

As the life cycle of the node is concentrated on delivering the required configuration and realization of the computing task, its following states may be distinguished (see Fig. 6.10):

- *offline*—the node has been started and the initialization has begun,
- *initialized*—the node has been properly initialized and it is waiting for the configuration, if the computing configuration is passed as a starting parameter the configuration loading begins,

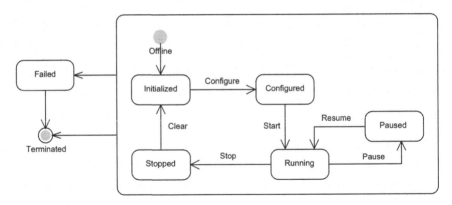

Fig. 6.10 Node life cycle

- *configured*—the computing task has been properly configured and the node is ready for starting the task; the computing task may be started on demand (e.g. after receiving a message from the user interface) or automatically, after loading the configuration,
- *running*—the computing is running,
- *paused*—the task has been paused and it is possible to re-run it again,,
- *stopped*—the computing task has stopped and the node can be safely turned off,
- *failed*—the node is in this state in any moment because of any unexpected failure,
- *terminated*—the node has finished working.

Figure 6.11 shows an example node with registered services. The figure distinguish the main service (called *core service*), which constitutes an execution environment for agents, that provides functionalities such as global addressing schema, communication via message passing, query mechanism, life-cycle management.

This service also plays role of a proxy between agents and other services. Various services provide functionalities related to infrastructure (e.g. communication and configuration provider services), simulation (e.g. stop condition service) or external tools (e.g. monitoring service, which collects and stores simulation data for a visualization application).

In one distributed environment particular nodes can have different responsibilities such as an end-user console, monitoring, management, and at last execution nodes. The role of a node is specified by the configuration of services plugged into its bus. Also, one can imagine a platform comprised only from a single node that works without any communication services.[6]

---

[6]Such configuration is often used for test purposes.

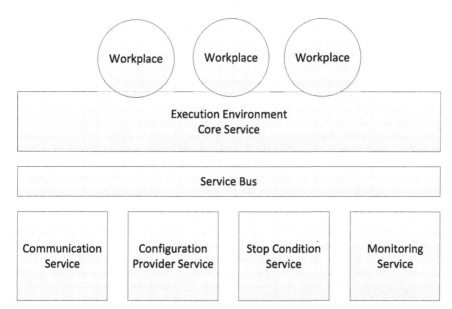

Fig. 6.11 Example node with services

6.3.2   Distributed Environment

The platform introduces *a virtual execution environment* in distributed systems that allows for performing operations involving top level agents (workplaces) located on different nodes without their awareness of physical distribution. Such operations are executed by the core service according to *Proxy* design pattern [124]. The service uses the communication service to communicate with core services located on other nodes. This constitutes *a global name space* of top level agents in the distributed environment. In other words, the virtual execution environment can be perceived as a realisation of *virtual root agent*.

The name space of agents can be narrowed by introducing *agent neighborhood* that defines visibility of top level agents. An agent can perform operations only on agents from its neighborhood. The neighborhood is realized and managed by a topology service (Fig. 6.12). This allows for creating virtual topologies among agents on the top of the distributed environment. Various topology strategies such as ring, grid, multi-dimensional grid can be applied in simulations.

Communication among the services located on different computing nodes is realized using dedicated communication protocol based on message passing. The communication is available by the ICommunicationService interface dedicated for sending messages and querying for incoming ones, addressed to the particular recipient. The communication between services is realized using the IIncomingMessageListener interface responsible for delivering the message

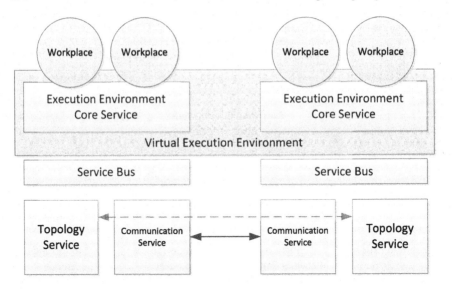

Fig. 6.12 Virtual execution environment

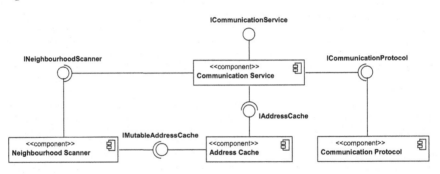

Fig. 6.13 Komponenty usugi komunikacyjnej

to the recipient. Current implementation is based on the four main elements shown in Fig. 6.13.

- *Communication Manager*—responsible for sending and receiving the messages, implements the interface of the communication service,
- *Neighbourhood Scanner*—responsible for detecting of other nodes of the platform, adds the addresses of the available nodes to the cache,
- *Communication Protocol*—responsible for physical sending of messages between the nodes, in the default implementation is based on Hazelcast library.
- *Address Cache*— stores the current addresses of other nodes that are delivered by the *Neighbourhood Scanner*.

6.3.3 Migration of the Agents

In the distributed environment it is assumed that the agents along with the dependent strategies may be migrated to different nodes, usually in order to use the computing power optimally or to follow some logical assumptions (e.g. individual migrates between the islands located on different nodes). The migration is realized by a dedicated service registered on each node. This service is responsible for serialization of agent (along with its dependencies), and later for its deserialization on the target node (so called *lightweight* migration). Migration services also use common communication service to exchange messages between the nodes.

The first step of the migration is serialization of the agent state along with its dependencies to so called state definition that describes current values of the agent properties and its dependencies (each of the migrated dependency has its own state definition). Next the created definitions are sent using the communication service to the target node and passed to the migration service. This service is responsible for recreation of the agent instance and passing it to the computing service (ICoreComponent, that places it in an appropriate place in the agent hierarchy (using the IoC container).

On the target node, during the recreation of the agent or its dependencies, there may arise a problem with lack of classes available to the virtual machine. In such case a process of remote loading of the required classes is run (so called *heavyweight* migration). This mechanism is realized using a dedicated *class loader*. Therefore remote class loading is completely transparent to the migration service. In the case of lack of certain class, (exception ClassNotFoundException) it is downloaded from a dedicated repository that is available on a certain node and loaded on the target node.

6.4 Summary

In this chapter AgE computing environment has been presented. This framework has been developed for over 15 years and since its inception was used multiple times in many applications, mostly connected with population-based computing, in particular the metaheuristics belonging to the class of multi-agent computing. The component oriented architecture of the platform makes it very easy to replace its different parts. Leveraging its reconfiguration properties leads to high usability in the aspect of implementation of different novel algorithms, as well as in the aspect of testing of their efficiency, monitoring the phenomena occurring in the populations etc. The research focused on AgE resulted later in creating many of its lightweight alternatives, using such languages as Scala, Python and recently Erlang. The research continues and the aspects of efficiency and flexibility of the platform remain the main issue for keeping in mind by the designers and developers of this solution.

Part III
Experimental Results

Chapter 7
EMAS in Optimization Problems

Considering the "no free lunch theorem" [297], it is still important to try to test how the examined metaheuristic works when applied to different well-known problems. That is why several well-known benchmark functions are considered in the following experimental study [82]. In order to obtain plausible results, the systems compared should be parametrized in the most similar way. So it is the case presented in this section, as EMAS, and its memetic variants (Baldwinian and Lamarckian) are compared with PEA (along with respective memetic modifications). Later, a comparison between EMAS and its immunological variant (iEMAS) is made. Finally two discrete problems are tackled: Low Autocorrelation Binary Sequence (LABS) and Optimal Golomb Ruler (OGR).

7.1 Continuous Optimization

7.1.1 Benchmarks Problems

The continuous benchmark functions considered in this monograph were selected from the set described in [82], their visualizations are presented in Fig. 7.1:

- Rastrigin:
  $f(x) = 10 \cdot n + \sum_{i=1}^{n}(x_i^2 - A\cos(2\pi x_i))$
  $-5.12 \le x_i \le 5.12$
  global minimum: $f(x) = 0$, $x_i = 0$, $i \in [1, n]$, see Fig. 7.1a.
- Ackley:
  $f(x) = -a \cdot e^{-b\frac{\sqrt{\sum_{i=1}^{n} x_i^2}}{n}} - e^{\frac{\sum_{i=1}^{n} \cos(c \cdot x_i)}{n}} + a + e;$
  $a = 20, b = 0.2, c = 2 \cdot \pi, i \in [1, n],$

© Springer International Publishing AG 2017
A. Byrski and M. Kisiel-Dorohinicki, *Evolutionary Multi-Agent Systems*,
Studies in Computational Intelligence 680, DOI 10.1007/978-3-319-51388-1_7

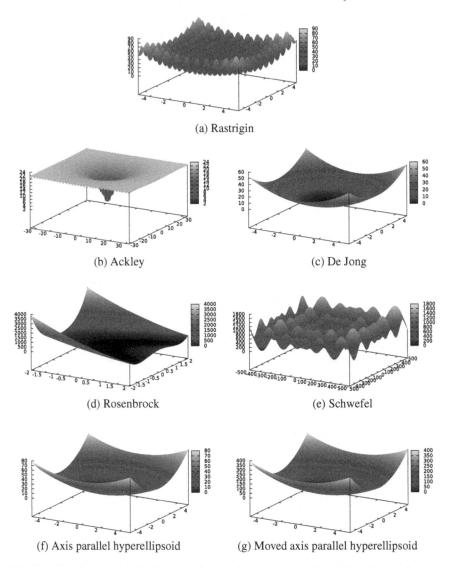

(a) Rastrigin

(b) Ackley   (c) De Jong

(d) Rosenbrock   (e) Schwefel

(f) Axis parallel hyperellipsoid   (g) Moved axis parallel hyperellipsoid

Fig. 7.1 Visualisation of the 2-dimensional cases of continuous benchmark functions used

$-1 \le x_i \le 1$,
global minimum: $f(x) = 0$, $x_i = 0$, $i \in [1, n]$, see Fig. 7.1b.

- De Jong:
$f(x) = \sum_{i=1}^{n} x_i^2$
$-5.12 \le x_i \le 5.12$,
global minimum: $f(x) = 0$, $x_i = 0$, $i \in [1, n]$, see Fig. 7.1c.

- Rosenbrock:
  $f(x) = \sum_{i=1}^{n-1} 100 \cdot (x_{i+1} - x_i^2)^2 + (1 - x_i)^2$
  $-2.048 \le x_i \le 2.048$,
  global minimum: $f(x) = 0$, $x_i = 0$, $i \in [1, n]$, see Fig. 7.1d.
- Schwefel:
  $f(x) = \sum_{i=1}^{n} -x_i \cdot sin(\sqrt{|x_i|})$
  $-500 \le x_i \le 500$,
  global minimum: $f(x) = -n \cdot 418.9829$, $x_i = 420.9678$, $i \in [1, n]$, see Fig. 7.1e.
- Axis Parallel Hyperellipsoid:
  $f(x) = \sum_{i=1}^{n} i \cdot x_i^2$
  $-5.12 \le x_i \le 5.12$,
  global minimum: $f(x) = 0$, $x_i = 0$, $i \in [1, n]$, see Fig. 7.1f.
- Moved Axis Parallel Hyper Ellipsoid:
  $f(x) = \sum_{i=1}^{n} 5 \cdot i \cdot x_i^2$
  $-5.12 \le x_i \le 5.12$,
  global minimum: $f(x) = 0$, $x_i = 5 \cdot i$, $i \in [1, n]$, see Fig. 7.1g.

All of the selected functions are multimodal or deceptive (where gradient information may lead the search astray), except for De Jong benchmark that is a convex function. This selection of course does not deplete the benchmarks, however, it seems to present a subjective, but reasonable set of reference problems, frequently used in testing population-based metaheuristics.

7.1.2 Classic EMAS and PEA

Here, before going to more sophisticated versions of these systems, the comparison results of classic EMAS with PEA are given. It is to note that PEA was implemented according to a scheme proposed by Michalewicz [210], namely population initialization, and looping through evaluation, selection, crossover and mutation of the individuals. In the discussed case real-valued encoding is used and allopatric speciation is introduced according to so-called island-model of evolution [54]. The problem considered in this section is optimization of 100-dimensional Rastrigin benchmark for several different configurations (after Pisarski, Rugała, Byrski and Kisiel-Dorohiniki [235]).

Common Configuration of the Tested Algorithms

- Representation: real-valued.
- Mutation: normal distribution-based modification of one randomly chosen gene.
- Crossover: single-point.
- Migration topology: 3 fully connected islands.
- Migration probability: 0.01 per agent/individual (each one migrates independently—possibly to different islands).

EMAS Configuration

- Initial energy: 100 units received by the agents in the beginning of their lives.
- Evaluation energy win/loose: 20 units passed from the looser to the winner.
- Minimal reproduction energy: 90 units required to reproduce.
- Death energy level: 0, such agents should be removed from the system.
- Boundary condition for the intra-island lattice: fixed, the agents cannot cross the borders.
- Intra-island neighborhood: Moore's, each agent's neighborhood consists of 8 surrounding cells.
- Size of 2-dimensional lattice as an environment: 10×10.
- Stop condition: 100000 steps of experiment.
- Benchmark problem: Rastrigin function [82].
- Problem size: 100 dimensions.
- Population size configurations:

    - 25 individuals on 1 island,
    - 25 individuals on 3 islands,
    - 40 individuals on 1 island,
    - 40 individuals on 3 islands.

Experimental Results

In Fig. 7.2, a simple comparison of the computation results, obtained for one-population configuration (1 island, 40 individuals), namely observation of the best fitness in each step of computation may be seen.

It turns out that EMAS outperforms PEA for over two orders of magnitude. This is a very promising result, and it may be further verified by checking the results obtained in both experiments in the last step presented in Table 7.1.

Fig. 7.2 EMAS versus PEA (1 island, 40 agents/individuals) *bestFitness(step)*

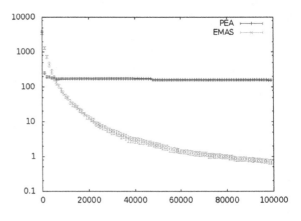

Table 7.1 EMAS and PEA optimization results obtained for 25 and 40 individuals

Number of islands	1 island		3 islands	
Computing system	EMAS	PEA	EMAS	PEA
25 individuals				
Result	1.77	242.96	0.71	156.34
St. Dev.	0.22	6.54	0.10	6.84
St. Dev. %	12.24	2.69	14.46	4.38
40 individuals				
Result	0.90	180.48	0.37	111.45
St. Dev.	0.13	6.45	0.05	4.16
St. Dev. %	15.24	3.57	13.56	3.73

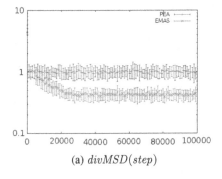
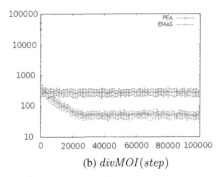

(a) *divMSD(step)*   (b) *divMOI(step)*

Fig. 7.3 Diversity of EMAS and PEA computed according to MOI and MSD schemes, 1 island 40 agents/individuals with standard deviation

In this table, the results obtained for other configurations (3 islands, 40 individuals) are also presented. In these cases, the domination of EMAS is still retained. It is easy to see that the results are repeatable (as standard deviation is relatively low).

Another important observation is that the multi-population models of computation tends to be better than single-population ones, as the former have better capabilities of exploration (as the population is decomposed to subpopulations), still retaining the capability of exploitation (in each single subpopulation). It should also be noted that increasing the number of individuals (from 25 to 40) improved the final result (although this requires a further proving and many more experimental runs).

In Fig. 7.3a, b, diversity computed according to MSD and MOI schemes are shown. It is easy to see that EMAS has lower diversity than PEA in both cases. However, it turns out that it does not hamper the computing efficiency (as EMAS outperforms PEA). Moreover, diversity, though lower, it is still quite stable (see standard deviation range marked on the graphs), which leads to the conclusion that though the population in EMAS is not so diverse as in PEA, the exploration and exploitation features of the system are balanced.

Table 7.2 EMAS and PEA MOI diversity for 25 and 40 individuals

Number of islands	1 island		3 islands	
Computing system	EMAS	PEA	EMAS	PEA
25 individuals				
Result	0.56	0.99	0.65	1.31
St. Dev.	0.22	0.14	0.29	0.14
St. Dev. %	38.67	14.04	44.05	10.87
40 individuals				
Result	0.50	1.00	0.57	1.30
St. Dev.	0.23	0.14	0.30	0.15
St. Dev. %	45.01	13.84	52.35	11.16

Table 7.3 EMAS and PEA MOI diversity for 25 and 40 individuals

Number of islands	1 island		3 islands	
Computing system	EMAS	PEA	EMAS	PEA
25 individuals				
Result	77.51	287.63	271.64	1253.43
St. Dev.	903.60	1004.92	2883.95	3000.84
St. Dev. %	1165.86	349.37	1061.69	239.41
40 individuals				
Result	135.22	313.53	460.94	1340.27
St. Dev.	1590.73	1030.57	4962.78	3066.69
St. Dev. %	1176.37	328.69	1076.66	228.81

The final values of both diversity measures (MOI and MSD) shown in Tables 7.2 and 7.3 confirm the observation of the computation behavior shown in Fig. 7.3a, b.

In order to examine the dynamics of the computing process for PEA and EMAS, an average number of steps between subsequent improvements of the best fitness observed are presented in Table 7.4. It is easy to see that also in this case, EMAS outperforms PEA. This result confirms that lower diversity of EMAS as compared to PEA does not hamper the capability of improving the result of the computation.

Another confirmation of the above observation may be found when looking at Table 7.5. There, a maximal number of steps needed for improving the value of the fitness function are shown. It is easy to see that EMAS outperforms PEA also in this case.

After checking that classical EMAS may become a reliable weapon of choice to deal with a difficult, e.g. black-box problem, it is time to consider other versions of EMAS and PEA capable of bringing substantial improvements to the computing process (e.g. balancing exploration and exploitation capabilities or reducing the number of fitness function calls).

Table 7.4 Average number of steps between subsequent improvements of the best fitness for 25 and 40 individuals

Number of islands	1 island		3 islands	
Computing system	EMAS	PEA	EMAS	PEA
25 individuals				
Result	116.90	133.74	142.61	195.33
St. Dev.	4.26	72.37	5.12	72.90
St. Dev. %	3.64	54.12	3.59	37.32
40 individuals				
Result	119.07	166.13	147.16	234.57
St. Dev.	4.43	87.15	5.97	77.02
St. Dev. %	3.72	52.46	4.05	32.84

Table 7.5 Maximal number of steps between subsequent improvements of the best fitness for 25 and 40 individuals

Number of islands	1 island		3 islands	
Computing system	EMAS	PEA	EMAS	PEA
25 individuals				
Result	4042.37	27441.57	3940.23	42225.17
St. Dev.	1580.74	17264.97	1330.81	17906.04
St. Dev. %	39.10	62.92	33.77	42.41
40 individuals				
Result	3850.93	33019.87	3596.30	47966.66
St. Dev.	1173.24	23218.82	889.32	20118.41
St. Dev. %	30.47	70.32	24.73	41.94

7.1.3 Memetic EMAS and PEA

The first and the most important thing to consider is the efficiency of the systems being compared, measured with the classical means (after Byrski, Korczyński and Kisiel-Dorohinicki [48]). Therefore, the fitness value of the best individual reached in certain generation (PEA) compared to a certain step of computation (EMAS) was examined. In Fig. 7.4, both PEA and EMAS fitnesses are shown, for all variants of these systems (evolutionary and memetic ones). The problem considered in this graph was 50-dimensional Rastrigin benchmark function.

It is easy to see that EMAS turns out to be generally better than PEA at evading the local extrema and continuing the exploration, while PEA is apparently already stuck. Drawing comparison between the memetic and evolutionary versions of the systems shows that Lamarckian memetics improves both PEA and EMAS, while

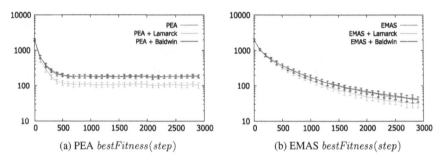

(a) PEA $bestFitness(step)$ (b) EMAS $bestFitness(step)$

Fig. 7.4 PEA and EMAS (memetic and evolutionary versions) fitness for 50-dimensional Rastrigin benchmark

Baldwinian memetics in the case considered produces results quite similar to the ones produced by the basic evolutionary versions.

Memetic Operators

It may be quite interesting to check how the application of particular memetic operators changes the outcomes of the experiments. In the given experiments, gradient-free steepest descent algorithm based on choosing the best one from 10 potential mutated individuals was used. Such a procedure was repeated 10 times and the best up-to-date result was returned. The results of the experiments are shown in Fig. 7.5.

The local search algorithms were implemented according to the two following strategies:

- Isotropic mutation—it is a method aimed at generating uniform sampling points on and within N-dimensional hyperspheres. The idea of the Isotropic method algorithm is as follows: firstly the N normal distributed numbers z_i are generated. Then the vectors x are computed by making a projection onto surface by dividing each generated number z_i by $r = \sqrt{\sum_{i=1}^{N} z_i^2}$. Since the z vectors are isotropically distributed, the vectors x will be of norm 1 and also isotropically distributed. There-

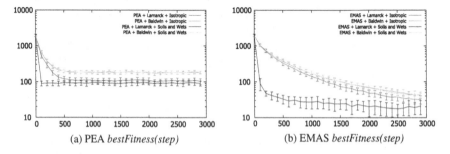

(a) PEA $bestFitness(step)$ (b) EMAS $bestFitness(step)$

Fig. 7.5 PEA and EMAS fitness for two different memetic operators (Solis Wets and isotropic mutation) applied

fore, the points will be distributed uniformly on the hypersphere. The generation of points inside the hypersphere may be achieved by rescaling the coordinates obtained in the previous steps [193].

- Solis and Wets' algorithm—it is a randomized version of optimization technique which belongs to a hill climbing family. Every step size can be adapted as follows: it starts at a current point $\mathbf{x}$ and checks if either $\mathbf{x} + \mathbf{d}$ or $\mathbf{x} - \mathbf{d}$ is better, where $\mathbf{d}$ is a deviate chosen from a normal distribution whose standard deviation is given by a parameter ρ. If the answer is positive then a move to a better point is made and a success is recorded. Otherwise, a failure is recorded. If several successes in a row happen, parameter ρ is increased to make longer moves. However, if several failures in a row are recorded, the ρ is decreased to focus the search [269].

It is easy to see that applying dedicated complex local search method (Solis-Wets operator) helps in quick approach to the final candidate solution for all the experiments. PEA gets stuck again in a local extremum, while EMAS retains better exploration capabilities which is pointed out by the curvature of the graphs, and also by higher standard deviation of the results. It is noteworthy that enhancing EMAS with dedicated local-search method (Solis-Wets), yielded very good results for Lamarckian memetics, while Baldwinian search again was the worst.

Number of Fitness Function Calls

The results observed at the beginning of this section, in particular PEA being outperformed by EMAS (see Fig. 7.4 and Table 7.7) require several important aspects to be discussed. First, it should be noted that in the comparison between these two algorithms, based on the number of steps or generations, EMAS is somewhat more handicapped. In EMAS, distributed selection mechanism allows for parallel ontogeny so that at one observable moment a certain number of individuals in the population are about to die, another group can be almost ready for reproduction etc. In PEA global synchronization is used and the whole population of individuals is processed at once. Therefore, PEA seems to be potentially better suited for exploration purposes, as it processes significantly more individuals than EMAS: conversely EMAS turns out to be more efficient.

Distributed selection mechanism in EMAS results in the fact that in one step of system work, the number of fitness function calls is far lower than in PEA. In Fig. 7.6, the number of new individuals produced in each step for PEA, EMAS and for their modifications is shown. It should be noted that in PEA, the number of fitness function calls per generation is constant and equals the number of individuals that was 90 per one generation (30 per one island) in the experiments. Lamarckian and Baldwinian modifications lead to multiplication of this number in the case of the experimental results by 100, therefore the number of fitness computations for memetic PEAs equals 900 per generation.

The number of fitness function calls in the case of EMAS, which oscillates around 10, is a significant advantage of this computing method. Moreover, Lamarckian modification of EMAS leads to obtaining about 200 fitness computations per step (still significantly lower than in the case of PEA), whereas Baldwinian is the most

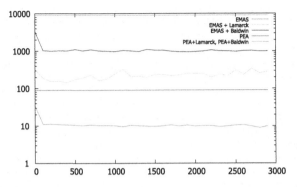

Fig. 7.6 Number of fitness function calls in EMAS and its memetic variations

Table 7.6 Execution time of steps

System	Avg. time [ms]	Std. dev	Std. dev %
EMAS	82.46	25.43	30
EMAS + Lamarck	138.48	45.03	32
EMAS + Baldwin	120.84	34.02	28
PEA	75.63	6.05	7
PEA + Lamarck	487.84	127.48	26
PEA + Baldwin	79.41	10.71	13

costly one with 1000 as estimate of the number of fitness computations. It is easy to see that the low number of fitness function calls for EMAS makes it an interesting weapon of choice for dealing with problems characterized by a costly fitness function (e.g. inverse problems). In this case, the complexity of the implementation of the whole system supporting the notion of agency, communication, naming services etc. is overwhelmed by the complexity of the fitness function. Therefore, looking for a more intelligent search algorithm becomes reasonable, despite the intrinsic complexities imposed by its implementation.

Step Execution Time

The most accurate evaluation of efficiency of the systems is observation of the execution time of computation, or as in the case of this analysis, of one computation step. Average times of step execution gathered during one experiment are shown in Table 7.6, along with dispersion estimation.[1]

At first glance, computing with EMAS bears higher time cost than with PEA, however one must remember that these results were collected for the case of optimization of simple benchmark function, and the total execution time will surely be

---

[1]These results were gathered using a server-class hardware (SUN FIRE X2100: Dual-Core AMD Opteron® Processor 1220 2.8 GHz, 4 GB RAM (2 × 2 GB), 1 × 250 GB SATA).

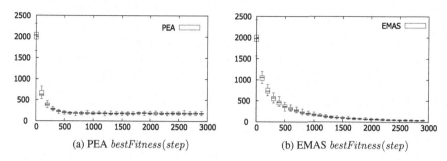

(a) PEA *bestFitness(step)* (b) EMAS *bestFitness(step)*

Fig. 7.7 Box-and-whiskers plot for EMAS and PEA fitness

much higher for PEA than for EMAS in the case of complex fitness function (cf. the paragraph describing the average number of fitness function calls in the current section). Moreover, these results clearly show that computing with EMAS is more unpredictable than with PEA, as the standard deviations are higher.

Experiments Repeatability

High dispersion of the results calls for additional analysis of repeatability of the experiments. Therefore, *box-and-whiskers* plots (containing minimum, maximum value, median, first and third quartiles) were prepared for selected experimental runs. In Fig. 7.7 these plots are presented for PEA and EMAS fitnesses. Based on standard deviation, it is easy to see that these experiments are repeatable.

Diversity

Observation of diversity measures for the systems show that the non-zero diversity is still retained, though falls down from the beginning of computation. However, the curvature of the graphs show that loss of diversity is much quicker in PEA than in EMAS (see Fig. 7.8). The same observation holds for all memetic variants of the system. Fortunately, it never reaches zero (compare experiments presented in [217]). Both diversity measures (MOI and MSD) yield generally similar results (loss of diversity is clearly seen, and EMAS losses diversity significantly later than PEA).

Dimensionality of the Problem

The degree of difficulty of the problems may be measured in different ways, considering e.g. the number of local extrema, features of the search space, dimensionality etc. In order to check, how the systems behave when they are faced with problems of different difficulty, dimensionality was chosen as its determinant. Therefore, an experiment based on conducting search for optima for benchmarks described in spaces of different dimensionality was conducted. It should be noted that such a test is useful to assess features of the search systems, as the high dimensionality of input and output variables presents an exponential difficulty (i.e., the effort grows exponentially with dimensions) for both problem modeling and optimization [179].

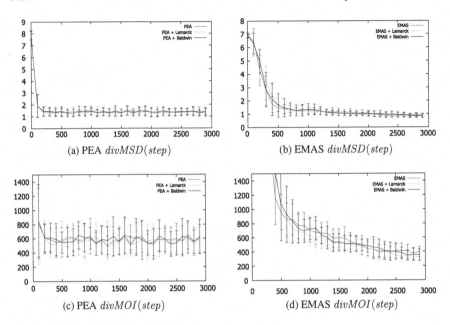

(a) PEA $divMSD(step)$

(b) EMAS $divMSD(step)$

(c) PEA $divMOI(step)$

(d) EMAS $divMOI(step)$

Fig. 7.8 PEA and EMAS diversity measurements

Figure 7.9 shows the graphs depicting the fitness value dependent on the simulation step for PEA and EMAS beginning from the 10-dimensional Rastrigin problem, and finishing on 100-dimensional one. It is easy to see that dimensionality of the problem greatly influences the efficiency of the techniques researched. It is easy to see that the higher dimension, the lower efficiency of the search technique.

Another important feature that may be found is the exploration capability. For PEA experiments, all the examined methods seem to get stuck in a local extremum, starting quite early (about the 1000-th generation in the case of higher dimensions, and even the 100-th generation when considering lower ones). In the case of EMAS techniques, a similar feature is a little bit harder to observe (besides the lowest dimensions) for both memetic operators used, and capabilities of improving the search result for PEA and EMAS using Lamarckian variation operators are not clearly visible.

Outcome of the Computation

In Table 7.7, the final values obtained in the 3000-th step are presented. It seems that EMAS outperformed all other systems (the best final results are pointed out using bold font). However, as the selected benchmarks are difficult, which is caused by its nature and high dimensionality, dispersion of the results is quite high. The diversity-related measures presented in Table 7.8 show that the final diversity is the highest in the case of PEA experiments. However (see Fig. 7.8), the curvature of the graphs point out that though the final result of the diversity for EMAS may be worse than

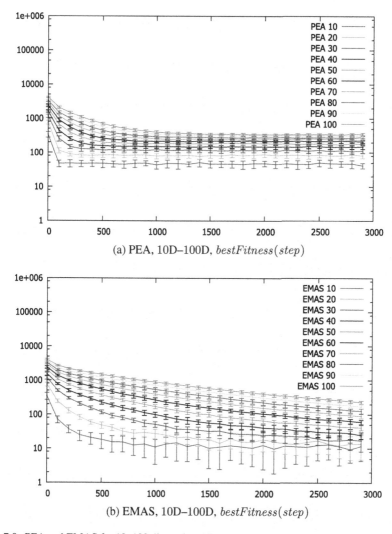

Fig. 7.9 PEA and EMAS for 10–100 dimensional Rastrigin problem

for PEA, the diversity is lost more slowly, helping to reach better final results than those obtained for PEA.

7.1.4 Classic and Immunological EMAS

Introduction of the immunological selection, which is in fact a distributed tabu list, leads to an important question, as to whether this mechanism hampers the efficiency

Table 7.7 The results obtained in the 3000-th step of computation for all the examined systems

	Result	St. Dev.	Min value	Median	Max value
Ackley					
PEA	2.03	0.16	1.66	2.09	2.26
EMAS	0.58	0.17	0.27	0.59	0.93
PEA + Lamarck	1.47	0.19	0.95	1.44	1.83
EMAS + Lamarck	**0.43**	0.20	0.09	0.44	0.85
PEA + Baldwin	2.03	0.14	1.77	2.03	2.38
EMAS + Baldwin	0.58	0.19	0.26	0.59	0.92
de Jong					
PEA	7.14	1.57	5.09	7.02	10.78
EMAS	0.81	0.65	0.18	0.49	2.79
PEA + Lamarck	4.34	1.20	2.25	4.03	7.17
EMAS + Lamarck	**0.67**	0.87	0.03	0.35	4.31
PEA + Baldwin	7.46	1.30	5.40	7.26	10.65
EMAS + Baldwin	0.70	0.60	0.18	0.56	3.17
Rosenbrock					
PEA	1219.37	346.69	741.61	1162.32	2478.04
EMAS	206.46	115.42	70.02	168.26	572.19
PEA + Lamarck	694.01	222.61	391.96	668.68	1557.13
EMAS + Lamarck	**190.49**	137.38	49.99	151.99	629.38
PEA + Baldwin	1233.87	285.51	766.65	1184.23	1760.46
EMAS + Baldwin	264.51	262.63	78.72	195.85	1266.85
Schwefel					
PEA	−12942.72	790.64	−14669.28	−12840.60	−11152.09
EMAS	**−13999.87**	1023.44	−15942.66	−14194.64	−11767.72
PEA + Lamarck	−13557.15	725.19	−14560.63	−13640.47	−11800.74
EMAS + Lamarck	−13680.38	901.95	−15517.86	−13682.55	−11695.06

(continued)

Table 7.7 (continued)

	Result	St. Dev.	Min value	Median	Max value
Axis parallel hyper ellipsoid					
PEA + Baldwin	−13051.08	795.88	−14274.21	−13157.16	−11370.22
EMAS + Baldwin	−13684.54	827.41	−15756.03	−13621.34	−12112.40
PEA	155.83	30.89	93.62	152.67	229.39
EMAS	16.11	15.05	3.46	9.96	63.12
PEA + Lamarck	82.56	26.84	22.06	80.32	146.14
EMAS + Lamarck	**9.90**	13.87	0.62	7.11	78.17
PEA + Baldwin	153.35	25.93	99.18	153.01	205.77
EMAS + Baldwin	18.51	17.24	3.69	11.29	66.82
Moved axis parallel hyper ellipsoid					
PEA	784.53	162.22	471.90	770.15	1170.59
EMAS	**50.69**	39.10	15.54	35.19	181.81
PEA + Lamarck	445.18	126.97	261.73	402.70	769.96
EMAS + Lamarck	58.69	64.73	3.12	32.81	273.17
PEA + Baldwin	780.84	166.14	493.92	775.91	1119.10
EMAS + Baldwin	78.09	65.54	15.46	45.96	218.38

of the system and whether it has any benefits at all? Indeed, despite first doubts, after observing the graphs presented in Fig. 7.10, it is easy to see that though the fitness seems to be a little worse in iEMAS than in EMAS, the number of agents is significantly lower during the whole computation. Therefore, this mechanism may produce interesting results, when the complexity of the fitness function is high and any means of decreasing the number of fitness function evaluation are crucial.

Experimental Configuration

In order to compare EMAS and iEMAS, 50-dimensional Rastrigin problem was used with the remaining parameters set as follows:

- Energy taken by a lymphocyte from similar agent: 30
- Good fitness factor: 0.97 (percentage of the agent fitness related to average fitness in the population, as minimization is considered, if fitness is smaller than average fitness, it is considered as "good").

Table 7.8 The diversity measures obtained in the 3000-th step of computation for all the examined systems

	MSD diversity	MSD diversity st. dev.	MOI diversity	MOI diversity st. dev.
Ackley				
PEA	0.70	0.15	173.02	40.47
EMAS	0.31	0.03	46.45	7.23
PEA + Lamarck	0.65	0.12	182.92	35.83
EMAS + Lamarck	0.32	0.05	43.62	7.32
PEA + Baldwin	**0.74**	0.16	**195.87**	42.56
EMAS + Baldwin	0.33	0.03	46.41	7.20
de Jong				
PEA	0.75	0.13	**254.35**	56.52
EMAS	0.52	0.11	74.19	14.37
PEA + Lamarck	0.72	0.10	226.00	39.86
EMAS + Lamarck	0.51	0.09	70.41	14.95
PEA + Baldwin	**0.77**	0.16	248.11	37.53
EMAS + Baldwin	0.54	0.10	78.35	15.45
Rosenbrock				
PEA	0.77	0.13	264.41	55.31
EMAS	0.51	0.12	81.17	23.04
PEA + Lamarck	0.69	0.10	232.09	34.17
EMAS + Lamarck	0.45	0.07	69.17	13.73
PEA + Baldwin	**0.81**	0.15	**276.26**	61.61
EMAS + Baldwin	0.49	0.09	73.06	14.87
Schwefel				
PEA	0.78	0.12	255.55	44.81
EMAS	1.82	0.42	1615.33	356.51
PEA + Lamarck	**6.67**	4.06	**3589.18**	2945.89
EMAS + Lamarck	1.69	0.29	1426.96	298.61
PEA + Baldwin	0.88	0.16	266.75	51.20
EMAS + Baldwin	1.81	0.37	1615.79	303.59
Axis parallel hyper ellipsoid				
PEA	**1.05**	0.38	298.34	73.49
EMAS	0.57	0.13	90.35	20.05
PEA + Lamarck	0.98	0.33	271.58	56.41

(continued)

Table 7.8 (continued)

	MSD diversity	MSD diversity st. dev.	MOI diversity	MOI diversity st. dev.
EMAS + Lamarck	0.56	0.13	81.67	17.92
PEA + Baldwin	1.04	0.23	**315.22**	71.53
EMAS + Baldwin	0.56	0.11	84.28	17.67
Moved axis parallel hyper ellipsoid				
PEA	0.94	0.18	297.35	57.34
EMAS	0.53	0.10	82.68	17.58
PEA + Lamarck	**1.08**	0.43	304.14	94.99
EMAS + Lamarck	0.53	0.09	78.85	15.87
PEA + Baldwin	1.07	0.33	**324.78**	77.17
EMAS + Baldwin	0.54	0.10	88.75	22.06

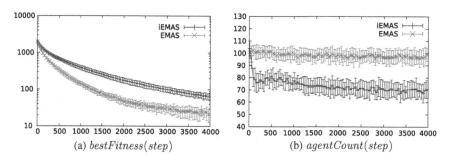

(a) $bestFitness(step)$ (b) $agentCount(step)$

Fig. 7.10 Fitness and agent count for EMAS and iEMAS

- Similarity measure: Mahalanobis distance [196].
- Similarity threshold: 7.3, if similarity is smaller than this the lymphocyte is considered to be similar to the tested agent.
- Immaturity duration for lymphocyte: 10.
- Maturity duration for lymphocyte: 20.
- Lymphocytes cannot migrate between the islands.

Experimental Results

The results presented in this section are recalled after Byrski [41].

The above observation gets confirmation when looking at Fig. 7.11, which shows the number of lymphocytes and the number of fitness function calls. Of course, as the number of agents is lower, it directly affects the number of fitness function calls, as predicted. Moreover, the number of lymphocytes is stable, therefore this feature

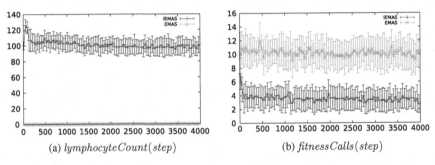

Fig. 7.11 Lymphocyte count and number of fitness function calls for EMAS and iEMAS

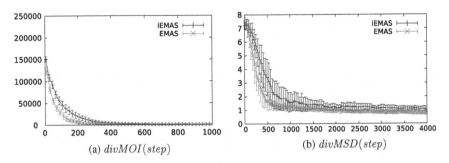

Fig. 7.12 MOI and MSD diversity for EMAS and iEMAS

of the computation may be easily predictable and adequate efforts may be made to tune the computing framework and the hardware to support an additional group of agents.

An interesting observation may be made after analyzing the graphs shown in Fig. 7.12. Both diversity measures point out that the diversity in iEMAS is a little better than in EMAS. This may result from the fact that the population of agents is affected by the lymphocytes. As similar agents are removed from the system (in fact, those similar to the ones recently removed), the diversity rises (this may be perceived as an effect similar to fitness sharing [198]).

7.2 Discrete Problems

7.2.1 Benchmarks Problems

In this section two problems (namely for LABS and Golomb Ruler) are described. These problems are difficult to solve, but quite easy to encompass in evolutionary environment, giving opportunity to explore in-depth search capabilities of the presented systems.

LABS

Low Autocorrelation Binary Sequence (LABS) is an NP-hard combinatorial problem with simple formulation. It has been under intensive study since 1960s by Physics and Artificial Intelligence communities. In consists in finding a binary sequence $S = \{s_0, s_1, \ldots, s_{L-1}\}$ with length L where $s_i \in \{-1, 1\}$ which minimizes energy function $E(S)$:

$$C_k(S) = \sum_{i=0}^{L-k-1} s_i s_{i+k} \quad E(S) = \sum_{k=1}^{L-1} C_k^2(S).$$

LABS has many applications in telecommunication (synchronization, pulse compression, satellite and space applications [125], digital signal processing, high-precision interplanetary radar measurements [287]), meteorology (calibration of surface profile meteorology tools [17]), physics (ising spin glasses, configuration state analysis, statistical mechanics [22]) and chemistry [22, 77, 271, 281, 282].

The search space for the problem with length L has size 2^L and energy of sequence can be computed in time $O(L^2)$. LABS problem has no constrains, so S can be represented naturally as an array of binary values [158]. In this problem all elements are correlated – there are no blocks of good solutions.

One of the reason of high complexity of problem is that in LABS all elements are correlated. One change that improves some $C_i(S)$, has also an impact on many other $C_j(S)$ and can lead to big changes of solution's energy.

The second problem is that, LABS has only few global optima for most values of L [123]. The search space is dominated by local optima. In [141] Halim compares search space to a "golf field", where GO are deep, isolated and are spreader just like "golf holes".

Golomb Ruler

A correct n-marks Golomb Ruler is an ordered, distinct, nonnegative, increasing sequence of integers $(g_1, g_2, \ldots, g_n)$ such that $g_i < g_{i+1}$ and all distances $g_j - g_i$ for $1 \leqslant i < j \leqslant n$ are unique. By convention the first element of the sequence g_1 equals 0 and the last element a_n is the length of the ruler.

Golomb Ruler was formulated by professor Solomon Wolf Golomb in 1977 [29]. It has various applications in radio communication (signal interference), coding theory, X-ray crystallography and radio astronomy [11, 29, 122, 200, 216, 242].

Finding the shortest n-marks Golomb ruler is challenging NP-hard combinator problem. The search space is tremendous and it grows exponentially with number of marks [261]. This is the main obstacle to discovering new rulers because $(n + 1)$–marks problem is much larger then n–marks problem.

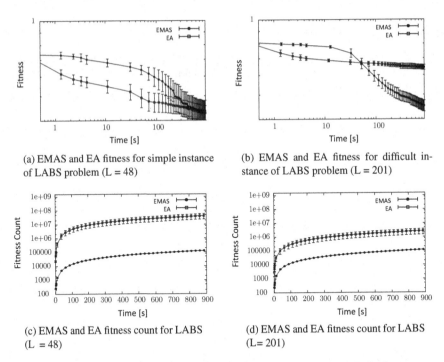

(a) EMAS and EA fitness for simple instance of LABS problem (L = 48)

(b) EMAS and EA fitness for difficult instance of LABS problem (L = 201)

(c) EMAS and EA fitness count for LABS (L = 48)

(d) EMAS and EA fitness count for LABS (L= 201)

Fig. 7.13 EMAS and EA comparison for simple and difficult instance of LABS problems

7.2.2 Classic EMAS and PEA

It is quite difficult to find a proper competitor for EMAS (as it is a one-of-kind system, directly hybridizing notions of agency with evolutionary computing), as The EA used for comparison followed Michalewicz model [13] and was selected as a relatively similar, general-purpose optimization algorithm, not utilizing any agent-oriented features. The parameters of both systems were made as similar, as it was possible (e.g. tournament selection was used for EA as it is very similar to meeting mechanism present in EMAS). In the cases of variation operators (crossover and mutation) the were of course retained completely the same for both systems. The efficiency was measured with regard to time, instead of system steps (as it is usually difficult to compare EMAS and PEA in this way, because one system step means something completely different in these two systems: in EA one step processes the whole generation, while in EMAS—one agent).

In EA and EMAS, tournament selection with tourney size of 2 was used. For LABS problem, uniform recombination was used, since it provides a new promising starting point for the algorithm. In Golomb Ruler problem, one point crossover was applied. In EA population size, chance to mutation and recombination probability are 50, 0.75, 0.5 for LABS and 100, 0.2, 0.8 for Golomb, respectively. In EMAS agents's count at the beginning, energy of reproduction, death and transfer are 50,

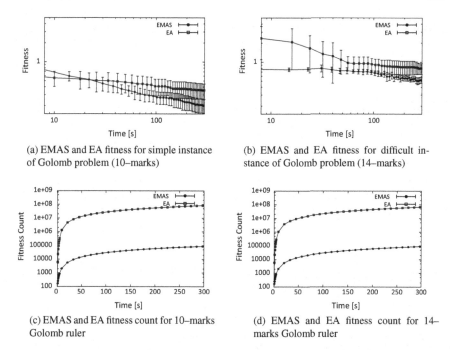

(a) EMAS and EA fitness for simple instance of Golomb problem (10–marks)

(b) EMAS and EA fitness for difficult instance of Golomb problem (14–marks)

(c) EMAS and EA fitness count for 10–marks Golomb ruler

(d) EMAS and EA fitness count for 14–marks Golomb ruler

Fig. 7.14 EMAS and EA comparison for simple and difficult instance of Golomb Ruler problems

45, 0, 5 for both problems. Each experiment was repeated 30 times with the same parameters and standard deviation was presented in the graphs in addition to the actual results. Total execution time is 900 s for LABS and 300 s for Golomb, and in fact the time was marked on X-axis (see Figs. 7.13 and 7.14). In figures and tables mean values of fitness with standard deviations of the best-so-far solution are given. The fitness results were scaled with respect to the instances of the problems tackled, and 0 means the best know optima (minimization assumed).

It took about 20 h of computations on a workstation with Intel Core 2 Quad Q8300 2.5 GHz (4 cores), 8 GB of DDR3 RAM, Windows 7 ×64 to get results for LABS and Golomb Ruler problems presented in this paper.

For simple instance of LABS problem presented in Fig. 7.13a EMAS turned out to be significantly worse at the beginning and close (though a little bit worse) at the end of computation. On the other hand, for significantly harder problem Fig. 7.13b EMAS is worse only at the beginning of computation. After some time EMAS evidently prevails over EA, it seems that the latter becomes stuck in a local extremum, while EMAS is still capable of exploring the search space (see final results presented in Table 7.9).

Note that EMAS calls fitness function significantly rarer than EA (see Fig. 7.13c, d, where aggregated counts of fitness function calls were presented). Based on the final results shown in Table 7.9, it is easy to compute, that computing with EMAS

Table 7.9 Final result

	EA	EMAS
(a) LABS $L = 48$		
Fitness	0.43 ± 0.03	0.44 ± 0.04
Fitness count	$4.2 \times 10^7 \pm 1.77 \times 10^7$	$1.3 \times 10^5 \pm 4.4 \times 10^3$
(b) LABS $L = 201$		
Fitness	0.64 ± 0.01	0.45 ± 0.02
Fitness count	$2.72 \times 10^6 \pm 1.21 \times 10^6$	$1.2 \times 10^5 \pm 2.93 \times 10^3$
(c) 10–marks Golomb Ruler		
Fitness	0.35 ± 0.10	0.20 ± 0.05
Fitness count	$8.47 \times 10^7 \pm 1.77 \times 10^7$	$8.7 \times 10^4 \pm 3.58 \times 10^3$
(d) 14–marks Golomb Ruler		
Fitness	0.80 ± 0.15	0.53 ± 0.05
Fitness count	$6.79 \times 10^7 \pm 8.25 \times 10^5$	$9.03 \times 10^4 \pm 1.28 \times 10^3$

requires 324 times less fitness function computation (for $L = 48$) and 22 times less ($L = 201$).

The results of experiments tackling Golomb Ruler problem show for both selected instances that EMAS becomes significant better that EA (see Fig. 7.14a, b for comparing the actual solutions obtained in the course of computation). EMAS still retains significantly lower cost of computing (in the means of fitness function count, see Fig. 7.14c, d). Again, final values obtained for both systems were shown in Table. 7.9.

Note that regardless of the number of fitness function calls, in three of the four conducted experiments, EMAS obtained significantly better results faster than EA (as it may be seen in the graps showing fitness value in relation to time of computation).

7.3 Summary

The presented chapter focused on testing the EMAS metaheuristic (its immunological and memetic variants) on several problems. In order to implement the simulations, Java and Python versions of the AgE platform were used. The problems tested belonged to the class of continuous and discrete optimization. The yielded results showed that EMAS has interesting properties and is a good competitor when taking into consideration classic evolutionary algorithm. In particular one of EMAS advantages is worth mentioning, namely the significant decrease of the number of fitness function calls, though reaching similar or better results than classic EA. Thus EMAS seems to be a weapon of choice for solving problems with complex fitness function evaluation.

Chapter 8
Tuning of EMAS Parameters

Having shown that EMAS approaches are effective in solving selected benchmark and real-life problems, it would be interesting to take an insight into the exact features of the most important mechanism of EMAS, i.e. the distributed selection based on existence of non-renewable resource. Such experiments could help to understand it and tune the computation based on this knowledge. The problem is not trivial, because EMAS, similar to other metaheuristics, utilises many parameters imposing on the user the setting dozens of degrees of freedom. The results presented in this section are recalled after Byrski [41].

Experimental Configuration

The configuration of the tested systems is presented as follows.

- Common parameters: normal distribution-based mutation of one randomly chosen gene, single-point crossover, the descendant gets parts of its parents genotype after dividing them in one randomly chosen point, 30 individuals located on each island, all experiments were repeated 30 times and standard deviation (or other statistical measures, such as median and appropriate quartiles for box-and-whiskers plots) was computed; allopatric speciation (island model), 3 fully connected islands, 3000 steps of experiment, genotype of length 50, agent/individual migration probability 0.01.
- PEA-only parameters: mating pool size equals to the number of individuals, individuals migrate independently (to different islands).
- EMAS-only parameters: initial energy: 100, received by the agents in the beginning of their lives, minimal reproduction energy: 90, required to reproduce, evaluation energy win/loose: 40/−40, passed from the looser to the winner, death energy level: 0, used to decide which agent should be removed from the system, boundary condition for the intra-island lattice: fixed, the agents cannot cross the borders, intra-island neighborhood: Moore's, each agent's neighborhood consists of 8 surrounding cells, size of 2-dimensional lattice as an environment: 10×10,

© Springer International Publishing AG 2017

A. Byrski and M. Kisiel-Dorohinicki, *Evolutionary Multi-Agent Systems*,

Studies in Computational Intelligence 680, DOI 10.1007/978-3-319-51388-1_8

all agents that decided to emigrate from one island, will immigrate to another island together (the same for all of them).

8.1 Energy-Related Parameters

Energy-based distributed selection mechanism is an immanent feature of EMAS. Therefore a detailed examination of its parameters is crucial for better understanding of the search process, and for being able to effectively tune them in order to adopt them to solving particular problems.

8.1.1 Energy Exchange Rate

The most crucial parameter of the distributed selection mechanism in EMAS is the rate of energy exchange between the meeting agents. The influence of changing this parameter on the fitness and agent count in the population is shown in Figs. 8.1 and 8.2.

Fig. 8.1 Influence of agent exchange energy on EMAS fitness *bestFitness(step)*

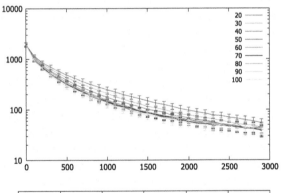

Fig. 8.2 Influence of agent exchange energy on EMAS agent count *agentCount(step)*

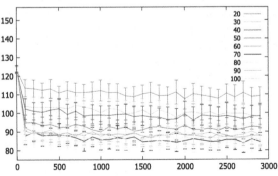

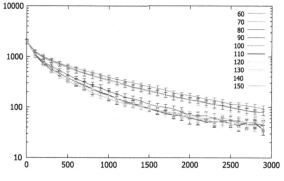

Fig. 8.3 Influence of agent initial energy on EMAS fitness *bestFitness*(*step*)

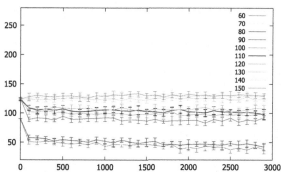

Fig. 8.4 Influence of agent initial energy on EMAS agent count *agentCount*(*step*)

It is easy to see that increasing this parameter makes the final result of computation better, but due to a logarithmic scale applied, this advantage does not seem to be significant. As predicted, this parameter greatly affects the agent count in the system. The higher the energy exchange rate, the lower the average agent count in the system.

8.1.2 Initial Energy Level

Influence of initial agent energy on the features of the system is presented in Figs. 8.3 and 8.4. Initial energy of the agents in the system is supposed to have a significant influence on the features of the agent population, as it is the main component of the total energy which is a base for distributed selection mechanism. In fact, looking at Fig. 8.3, the influence seems to be strong and straightforward.

The higher initial energy, the greater the number of agents during the computation. It should be noted that the selection mechanism is stable, as the number of agent does not grow indefinitely nor does it fall to zero during the whole computing process. It is easy to see that changing the initial energy affects indirectly the fitness in the system (see Fig. 8.4), changing the actual number of the agents in the system that are

Fig. 8.5 Influence of minimum reproduction energy on EMAS agent count *agentCount*(*step*)

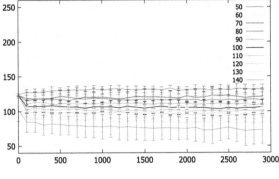

Fig. 8.6 Influence of minimum reproduction energy on EMAS fitness *bestFitness*(*step*)

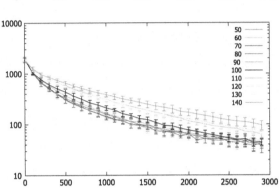

capable of exploring and exploiting the search space. Generally speaking, increasing the initial energy helps to reach better results, though this effect is not very apparent.

8.1.3 Minimal Reproduction Energy

Influence of minimal reproduction energy on the features of the system is presented in Figs. 8.5 and 8.6.

Minimal reproduction energy of the agents is supposed to have a significant influence on the features of the agent population, as it directly affects the distributed selection mechanism by controlling "maturity" of agents capable of reproduction. If this parameter value is low agents that performed few rendezvous will reproduce, while for its high value only the agents which live longer may generate offspring. In fact, looking at Fig. 8.6, the influence seems to be strong and straightforward, just opposite to the case of initial energy.

It is easy to see that the higher the minimal reproduction level, the lower the number of the agents during the computation, as it is more difficult for them to reproduce. Again, the selection mechanism is stable, as the number of agent does not grow indefinitely nor does it fall to zero, during the whole computing process.

The fitness is also affected (see Fig. 8.5) because the number of the agents varies for different values of the minimal reproduction energy. The system is able to better and quicker explore the search space for lower levels of this parameter (final results of the search are better for lower values of minimal reproduction energy, and the search is quicker as the graph curvature is higher).

8.2 Probabilistic Decision Parameters

Stochastic nature of the systems brings flexibility into the computing, but if EMAS and related techniques are to be used effectively, a detailed examination of the most important probabilistic decision parameters is necessary.

8.2.1 Migration Probability

The existence of migration phenomenon between the subpopulations should affect positively the value of fitness. It seems to be straightforward because such techniques as niching and speciation are meant to increase the exploration efficiency of the algorithm. Indeed, it is easy to see that introducing migration into the system enhances the quality of results (see Fig. 8.7), despite the fact effect is almost discrete—if the probability is non-zero, the results are significantly better, however, increasing of this parameter does not produce distinguishable changes in the fitness value. This may result from the fact that the evolutionary islands were fully connected; perhaps introducing more sophisticated topology would relax the influence of this parameter on the overall efficiency of the computation.

8.2.2 Meeting Probability

This parameter affects the frequency of meetings between the agents (as the decision whether or not the agent meets another agent is based on the outcome of probabilistic sampling). The higher the meeting probability is, the more frequently agents will meet and exchange their energy.

However, this parameter does not influence the number of agents in the population (see Fig. 8.8) because the same number of agents simply exchanges the energy faster or slower, also in the memetic versions of EMAS. Again the selection mechanism is stable, as the number of agent does not grow indefinitely nor does it fall to zero during the whole computing process.

Increasing the meeting probability makes it possible to reach the desired solutions quicker (see Fig. 8.9), as the energy flow from "worse" agents to "better" ones is faster, so "better" agents may reproduce quicker. Therefore, the final results of the search

Fig. 8.7 Influence of migration probability on EMAS fitness *bestFitness(step)*

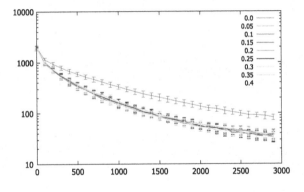

Fig. 8.8 Influence of meeting probability on EMAS agent count *agentCount(step)*

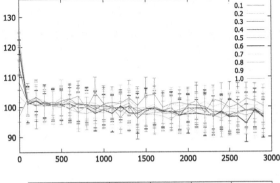

Fig. 8.9 Influence of meeting probability on EMAS fitness *bestFitness(step)*

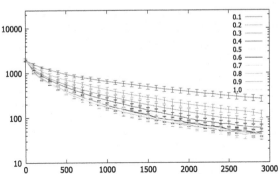

are better for higher values of meeting probability, and the search is quicker as the graph curvature is higher. Again, changing this parameter, does not greatly affect memetic modifications of EMAS.

Very important information may be obtained when observing diversity shown in Figs. 8.10 and 8.11. Increasing the meeting probability decreases diversity. As the presence of a diverse population is an important thing in the population-based search [54], one should choose the value of this parameter in such a way that the desired solution is approached as quickly as planned (as a result of exploitation), and diversity is high enough to maintain the exploration. Choosing an appropriate value

Fig. 8.10 Influence of meeting probability on EMAS MSD diversity *divMSD(step)*

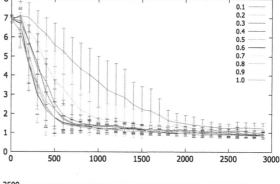

Fig. 8.11 Influence of meeting probability on EMAS MOI diversity *divMOI(step)*

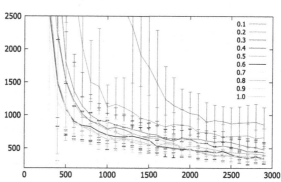

of this parameter seems to be crucial to maintain balance between exploration and exploitation for EMAS and its variations.

8.2.3 Immunological Parameters

As in Sect. 7.4, immunological variant of EMAS (iEMAS) is an important weapon of choice when dealing with problems which have a complex fitness function. Therefore, an examination of selected parameters influencing the immunological selection is necessary.

Penalty Threshold

This parameter may be described as a quantity of energy taken from the agent, which turns out to be similar to a lymphocyte during affinity testing. It is easy to see that changing this parameter significantly influences the number of agents in the system and yet the fitness remains almost unchanged (see Fig. 8.12a), which is a very interesting fact.

This observation clearly indicates that introducing distributed tabu mechanism defined in this way does not hamper the search capabilities of the system. Of course,

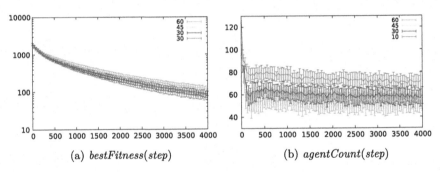

Fig. 8.12 Influence of penalty threshold on fitness and agent count in iEMAS

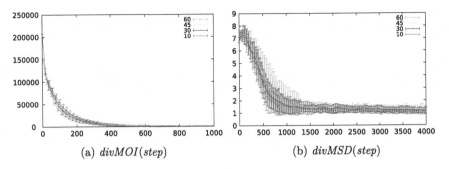

Fig. 8.13 Influence of penalty threshold on diversity in iEMAS

the higher penalty is, the more agents are removed from the system, therefore, the relation shown in (Fig. 8.12b) is predictable.

Observation of the diversity measures (see Fig. 8.13) shows that changing the penalty threshold (at the same time changing the immunological selection pressure) does not hamper the diversity. Moreover, as reported in Sect. 7.4, quicker removing of "bad" agents makes the system more diverse (in terms of MSD metric).

Penalty threshold has also a predictable influence on the number of lymphocytes in the system (see Fig. 8.14) closely connected with reducing of the agent population.

When the number of agents is lower, the same total sum of energy is distributed among individuals of a smaller population, therefore, the average value of energy per agent is higher and agents do not die so often as in the case of a bigger population. In effect, when the number of individuals in the population is low, also the smaller number of lymphocytes is generated.

Lymphocyte Life Length

Longer lymphocyte life (see Fig. 8.15) again does not significantly worsen the fitness, however certain influence may be observed, as the fitness becomes little better in the case of a shorter lymphocyte life. At the same time, of course, the agent count decreases with the rise of lymphocyte life as the lymphocytes may act longer removing the individuals from the population.

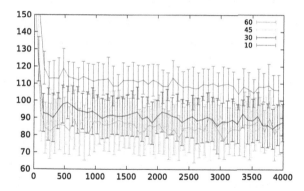

Fig. 8.14 Influence of penalty threshold on lymphocyte count

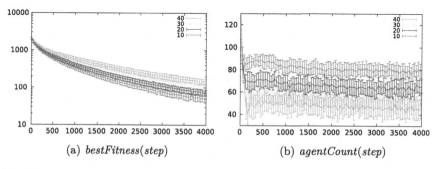

(a) *bestFitness*(*step*)   (b) *agentCount*(*step*)

Fig. 8.15 Influence of lymphocytes' lifespan on fitness and agent count

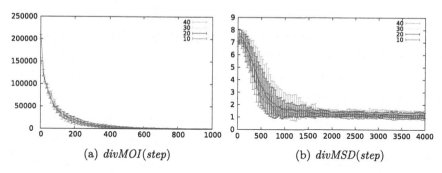

(a) *divMOI*(*step*)   (b) *divMSD*(*step*)

Fig. 8.16 Influence of lymphocytes' life length on diversity

At the same time, manipulating the lifespan of lymphocytes does not hamper the diversity measures, though a little positive influence may be observed in the case of MSD diversity, when the lymphocyte lifespan is longer (see Fig. 8.16).

It is interesting that the length of lymphocyte lifespan does not affect at all the number of lymphocytes in the system (see Fig. 8.17). It shows that the immunological selection mechanism is stable and lymphocytes do not tend to overpopulate agents,

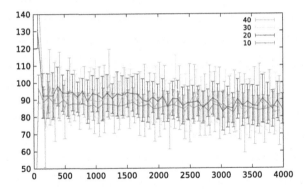

Fig. 8.17 Influence of lymphocytes' life length on lymphocyte count

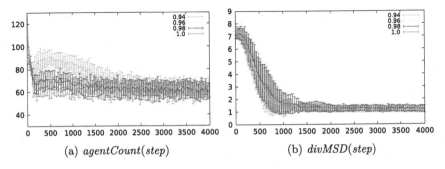

Fig. 8.18 Influence of "good" fitness percentage on agent count and diversity

though the average number count has a significant diversity, due to full stochastic nature of the selection mechanism.

Percentage of "Good" Fitness

During a negative selection process, lymphocytes are removed when they are still considered immature, though they match a "good" agent in the population. This is the case when an immature lymphocyte matches an agent that has fitness related in some way to the average fitness in the population (an appropriate percentage is considered). In Fig. 8.18 the results of changing this percentage are shown, along with the MSD diversity of the population.

It is easy to see that these two graph sets are related. When the population is diverse (mostly at the beginning of the computation) the level of "good" fitness is lower than later when the diversity falls down. So, lymphocytes tend to be removed more often, and therefore the population of agents is larger. Other important parameters such as fitness, MOI diversity and lymphocyte count are quite similar to those discussed before and remain unchanged in the relation with a given parameter.

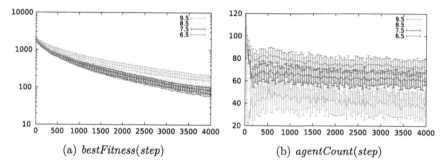

Fig. 8.19 Influence of similarity threshold on fitness and agent count

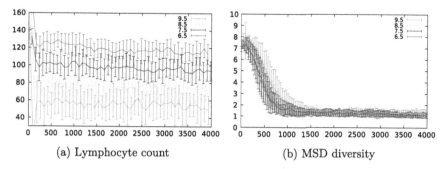

Fig. 8.20 Influence of similarity threshold on lymphocyte count and MSD diversity

Affinity Measure

In order to measure affinity between lymphocytes and agents, Mahalanobis distance was used [196]. Lower distance means that the lymphocyte must match a closer agent before penalizing it (and vice versa). Therefore, it is easy to see that increasing the distance hampers a little the obtained fitness, and of course decreases the number of agents in the system (see Fig. 8.19).

At the same time, observation of the lymphocyte count reveals that if the distance is lower, more lymphocytes are created in the system, as it is easier to remove the agent. It is connected of course with the similarity measure (this effect has been already observed before) that removing lymphocytes increases the MSD similarity measure (see Fig. 8.20).

8.3 Summary

The above-mentioned experiments may surely provide a basis for researchers who are willing to apply the EMAS-like computing to their problems. In order to make this easier, a summary of the parameters tuning is presented in Table 8.1. Based on the results presented in this table, in order to appropriately parametrize computation,

Table 8.1 Parameters tuning summary

Increase of the parameter	Fitness	Agent count	MOI	MSD	Lymphocyte count
Energy exchange rate	↗	↘	↗	—	
Initial energy level	↗	↗	—	—	
Minimal reproduction energy	↘	↘	↘	—	
Migration probability	↗	—	↗	—	
Meeting probability	↗	—	↘	↘	
Penalty level	—	↘	—	↗	↘
Good fitness percentage	—	↘	—	—	—
Lymphocyte life length	↘	↘	—	↗	—
Similarity distance	↘	↘	—	↗	↘

one must focus not only on attaining a specified goal (e.g. good fitness) but also check whether other parameters comply with this goal.

Therefore a desire to increase one parameter of the computation (e.g., fitness), must be verified with other criteria, for example a need to retain relatively small population of the agents or high diversity.

Synthetic results as these presented in Table 8.1 will surely help in such activities, creating a starting point for continuing of the research. That is, after finding interesting parameters interdependencies, the researcher should refer to one of the previous sections, describing the details of the experiments, and start to prepare and test his own system configurations.

Final Remarks

Solving difficult search and optimization problems (e.g. "black-box") requires non-deterministic approaches, in particular metaheuristics. It may be said that these methods trade-off precision, quality, accuracy and execution time in favor of computational effort. Such methods usually referred to as the *methods of last resort* are necessary for dealing with difficult problems.

Following *no free lunch theorem* mentioned earlier, it turns out that looking for novel metaheuristics for particular problems will always be necessary. Therefore, the existing metaheuristics are proposed to be enhanced with agency, constituting EMAS (proposed by Cetnarowicz) and its modifications (memetic and immunological), which yielded original results presented in this monograph. Moreover, it should be noted that developing novel computing methods calls for the development of a detailed formal model that will ease the understanding of the method, and provide a base for further analytical deliberations, helpful in getting a deeper insight into the features of the system. Accurate testing of the proposed metaheuristics is also necessary, in particular tuning the system parameters may be fruitful for further application of this metaheuristics to other problems.

Significant attention should also be devoted to the fact of appropriate modeling, design and development of dedicated software platforms, capable of efficiently and flexibly supporting the process of construction and using of the computational agent-based systems. Thus, one of the starting point for this monograph was the idea of connecting the possibilities given by the agent-based approach and component techniques in the field of population-based techniques of computational intelligence. The so-far constructed platforms were leveraging different technologies (like e.g. Java, Python, C#.NET), and allowed for verification of many technological methods, showing that efficient realization of such systems requires dedicated tools, which are flexible and extensible both on the configuration level and particular implementation parts of single agents.

Summing up, this monograph gives a guidance in the field of agent-based computing, in particular connected with design, evaluate and development of agent-based general-purpose metaheuristics for optimization problems. Besides presentation of the computing paradigm, references to the state-of-the-art and technological deliber-

© Springer International Publishing AG 2017

A. Byrski and M. Kisiel-Dorohinicki, *Evolutionary Multi-Agent Systems*,
Studies in Computational Intelligence 680, DOI 10.1007/978-3-319-51388-1

ations and an extensive set of experiments, a full formal model proving the ergodicity of a dedicated Markov chain is presented, in order to justify the reason for tackling such complex computing systems. Moreover, EMAS is treated as a starting point for further hybridization and extensions, so in this way memetic EMAS hybrids and immunological EMAS are discussed, which have already proved to be general-purpose search algorithms.

The authors continue with the research related to agent-based computing, focusing on novel optimization algorithms, exploring possibilities of using of heterogeneous hardware and new implementation technologies (like functional programming) in order to develop parallel and distributed versions of the computing platforms, capable of running on supercomputer machines.

References

1. *Inteligencia Artificial, Special issue: New trends on Multiagent systems and soft computing*, 9(28), 2005.
2. J. T. Alander. *An Indexed Bibliography of Genetic Algorithms: Years 1957–1993*. Art of CAD ltd, Vaasa, Finland, 1994.
3. E. Alba and M. Tomassini. Parallelism and evolutionary algorithms. *IEEE Trans. on Evolutionary Computation*, 6(5):443–462, 2002.
4. M. Ali, C. Storey, and A. Törn. Application of stochastic global optimization algorithms to practical problems. *Journal of Optimization Theory and Applications*, 95:545–563, 1997.
5. G. Almasi and A. Gottlieb. *Highly Parallel Computing*. Benjamin-Cummings Publishers, 1989.
6. D. Alur, J. Crupi, and D. Malks. *Core J2EE Patterns: Best Practices and Design Strategies*. Prentice Hall, 2003.
7. S. An, S. Yang, S. Ho, T. Li, and W. Fu. A modified tabu search method applied to inverse problems. *Magnetics, IEEE Transactions on*, 47(5):1234 –1237, may 2011.
8. J. Arabas. *Wykłady z algorytmów ewolucyjnych (in Polish, Lectures on Evolutionary Algorithms)*. WNT, 2001.
9. R. Aster, B. Borchers, and C. Thurber. *Parameter Estimation and Inverse Problems*. Elsevier, 2012.
10. M. Atallah. *Algorithms and Theory of Computation Handbook*. CRC Press LLC, 1999.
11. W. C. Babcock. Intermodulation interference in radio systems. *Bell Systems Technical Journal*, 32:63–73, 1953.
12. T. Bäck. *Evolutionary Algorithms in Theory and Practice*. Oxford University Press, 1996.
13. T. Bäck, D. Fogel, and Z. Michalewicz, editors. *Handbook of Evolutionary Computation*. IOP Publishing and Oxford University Press, 1997.
14. T. Bäck, U. Hammel, and H.-P. Schwefel. Evolutionary computation: Comments on the history and current state. *IEEE Trans. on Evolutionary Computation*, 1(1):3–17, 1997.
15. T. Bäck, U. Hammel, and H.-P. Schwefel. Evolutionary computation: Comments on the history and current state. *IEEE Trans. on Evolutionary Computation*, 1(1), 1997.
16. J. Baldwin. A new factor in evolution. *American Naturalist*, 30:441–451, 1896.
17. S. K. Barber, P. Soldate, E. H. Anderson, R. Cambie, W. R. McKinney, P. Z. Takacs, D. L. Voronov, and V. V. Yashchuk. Development of pseudorandom binary arrays for calibration of surface profile metrology tools. *Journal of Vacuum Science & Technology B: Microelectronics and Nanometer Structures*, 27(6):3213–3219, 2009.

© Springer International Publishing AG 2017
A. Byrski and M. Kisiel-Dorohinicki, *Evolutionary Multi-Agent Systems*,
Studies in Computational Intelligence 680, DOI 10.1007/978-3-319-51388-1

18. F. Bellifemine, G. Caire, A. Poggi, and G. Rimassa. JADE: A software framework for developing multi-agent applications. Lessons learned. *Information and Software Technology*, 50(1-2):10–21, 2008.

19. F. Bellifemine, G. Rimassa, and A. Poggi. JADE – A FIPA-compliant agent framework. In *Proc. of 4th Int. Conf. and Exhibition on the Practical Application of Intelligent Agents and Multi-Agents (PAAM'99)*, pages 97–108, 1999.

20. F. Bergenti, M. Gleizes, and F. Zambonelli. *Methodologies and Software Engineering for Agent Systems*. Kluwer Academic Publisher, 2004.

21. F. Bergenti, M. P. Gleizes, and F. Zambonelli. *Methodologies and Software Engineering for Agent Systems*. Kluwer Academic Publisher, 2004.

22. J. Bernasconi. Low autocorrelation binary sequences: statistical mechanics and configuration space analysis. *Journal de Physique*, 48(4):559–567, 1987.

23. H. Bersini and F. Varela. Hints for adaptive problem solving gleaned from immune networks. In H.-P. Schwefel and R. Mnner, editors, *Parallel Problem Solving from Nature*, volume 496 of *Lecture Notes in Computer Science*, pages 343–354. Springer Berlin Heidelberg, 1991.

24. N. Beume, M. Laumanns, and G. Rudolph. Convergence rates of (1+1) evolutionary multi-objective optimization algorithms. In R. Schaefer, C. Cotta, J. Kolodziej, and G. Rudolph, editors, *Proc. of 11th Int. Conf. on Parallel Problem Solving from Nature – PPSN XI*, volume 6238 of *LNCS*, pages 597–606. Springer, 2010.

25. N. Beume, M. Laumanns, and G. Rudolph. Convergence rates of (1+1) evolutionary multi-objective optimization algorithms. In R. Schaefer, C. Cotta, J. Kolodziej, and G. Rudolph, editors, *PPSN (1)*, volume 6238 of *Lecture Notes in Computer Science*, pages 597–606. Springer, 2010.

26. J. Bhasker. *A SystemC Primer, 2nd Edition*. Star Galaxy Publishing, 2004.

27. P. Biegański, A. Byrski, and M. Kisiel-Dorohinicki. Mulit-agent platform for distributed soft computing. In *Inteligencia Artificial, Special issue: New trends on Multiagent systems and soft computing* [1], pages 63–70.

28. P. Billingsley. *Probability and Measure*. John Wiley and Sons, 1987.

29. G. S. Bloom and S. W. Golomb. Applications of numbered undirected graphs. *Proceedings of the IEEE*, 65(4):562–570, 1977.

30. C. Blum and A. Roli. Metaheuristics in combinatorial optimization: Overview and conceptual comparison. *ACM Computing Surveys*, 35(3):268–308, 2003.

31. P. Bonissone. Soft computing: the convergence of emerging reasoning technologies. *Soft Computing*, 1(1):6–18, 1997.

32. P. Bouvry, H. González-Vélez, and J. Kołodziej. *Intelligent Decision Systems in Large-Scale Distributed Environments*. Springer, 2011.

33. G. E. P. Box. Evolutionary operation: A method for increasing industrial productivity. *Applied Statistics*, VI(2), 1957.

34. M. Bratman. *Intentions, Plans and Practical reason*. Harvard University Press, 1987.

35. R. Brooks. A robust layered control system for a mobile robot. *IEEE Journal of Robotics and Automation*, 2(1):14–23, 1986.

36. J. Brownlee. *Clever Algorithms: Nature-Inspired Programming Recipes*. lulu.com, 2012.

37. E. Burke, G. Kendall, J. Newall, E. Hart, P. Ross, and S. Schulenburg. Hyper-heuristics: An emerging direction in modern search technology. In F. Glover and G. Kochenberger, editors, *Handbook of Metaheuristics*, volume 57 of *International Series in Operations Research & Management Science*, pages 457–474. Springer US, 2003.

38. F. Burnet. *The clonal selection theory of acquired immunity*. Vanderbilt University Press, 1959.

39. A. Byrski. *Immunological Selection Mechanism in Agent-based Evolutionary Computation (in Polish: Immunologiczny mechanizm selekcji w agentowych obliczeniach ewolucyjnych)*. PhD thesis, AGH University of Science and Technology, 2007.

40. A. Byrski. *Agent-based metaheuristics in search and optimization*. Dissertations and monographs vol. 268, AGH University of Science and Technology Press, 2013.

41. A. Byrski. Tuning of agent-based computing. *Computer Science (accepted)*, 2013.

42. A. Byrski, R. Debski, and M. Kisiel-Dorohinicki. Agent-based computing in an augmented cloud environment. *Computer Systems Science and Engineering*, 27(1), 2012.
43. A. Byrski, R. Dreżewski, L. Siwik, and M. Kisiel-Dorohinicki. Evolutionary multi-agent systems. *The Knowledge Engineering Review*, 2013 (accepted for printing).
44. A. Byrski and M. Kisiel-Dorohinicki. Immunological selection mechanism in agent-based evolutionary computation. In M. A. Klopotek, S. T. Wierzchon, and K. Trojanowski, editors, *Intelligent Information Processing and Web Mining: proceedings of the international IIS: IIPWM '05 conference: Gdansk, Poland*, Advances in Soft Computing, pages 411–415. Springer Verlag, 2005.
45. A. Byrski and M. Kisiel-Dorohinicki. Agent-based evolutionary and immunological optimization. In *Computational Science – ICCS 2007, 7th International Conference, Beijing, China, May 27–30, 2007, Proceedings*. Springer, 2007.
46. A. Byrski, M. Kisiel-Dorohinicki, and E. Nawarecki. Agent-based evolution of neural network architecture. In M. Hamza, editor, *Proc. of the IASTED Int. Symp.: Applied Informatics*. IASTED/ACTA Press, 2002.
47. A. Byrski, M. Kisiel-Dorohinicki, and E. Nawarecki. Immunological selection in agent-based optimization of neural network parameters. In *Proc. of the 5th Atlantic Web Intelligent Conference AWIC'2007: Fontainebleau, France, June 25–27, 2007*, pages 62–67. Springer Advances in Soft Computing 43, 2007.
48. A. Byrski, W. Korczynski, and M. Kisiel-Dorohinicki. Memetic multi-agent computing in difficult continuous optimisation. In D. Barbucha, M. T. Le, R. J. Howlett, and L. C. Jain, editors, *Advanced Methods and Technologies for Agent and Multi-Agent Systems, Proceedings of the 7th KES Conference on Agent and Multi-Agent Systems - Technologies and Applications (KES-AMSTA 2013), May 27-29, 2013, Hue City, Vietnam*, volume 252 of *Frontiers in Artificial Intelligence and Applications*, pages 181–190. IOS Press, 2013.
49. A. Byrski, W. Korczyński, and M. Kisiel-Dorohinicki. Memetic multi-agent computing in difficult continuous optimisation. In *Proc. of 6th Int. KES Conf. on Agents and Multi-agent Systems Technologies and Applications*. IOS Press, 2013 (accepted for printing).
50. A. Byrski and R. Schaefer. Stochastic model of evolutionary and immunological multi-agent systems: Mutually exclusive actions. *Fundamenta Informaticae*, 95(2–3):263–285, 2009.
51. A. Byrski, R. Schaefer, and M. Smołka. Asymptotic guarantee of success for multi-agent memetic systems. *Bulletin of the Polish Academy of Sciences—Technical Sciences*, 61(1), 2013.
52. A. Byrski, R. Schaefer, and M. Smołka. Markov chain based analysis of agent-based immunological system. *Transactions on Computational Collective Intelligence*, 10, 2013.
53. A. Byrski, L. Siwik, and M. Kisiel-Dorohinicki. Designing population-structured evolutionary computation systems. In T. Burczyński, W. Cholewa, and W. Moczulski, editors, *Methods of Artificial Intelligence (AI-METH 2003)*, pages 91–96. Silesian University of Technology, Gliwice, Poland, 2003.
54. E. Cantú-Paz. A summary of research on parallel genetic algorithms. *IlliGAL Report No. 95007. University of Illinois*, 1995.
55. E. Cantú-Paz. *Efficient and Accurate Parallel Genetic Algorithms*, volume 1 of *Genetic Algorithms and Evolutionary Computation*. Springer Verlag, 2000.
56. Y. Cao and Q. Wu. Convergence analysis of adaptive genetic algorithm. In *Genetic Algorithms in Engineering Systems: Innovations and Applications (GALESIA 1997)*, pages 85–89, 1997.
57. E. Cardozo, J. Sichman, and Y. Damazeau. Using the active object model to implement multi-agent systems. In *Proc. of 5th IEEE Int. Conf. on Tools with Artificial Intelligence (TAI'93)*, 1993.
58. K. Cetnarowicz. Evolution in multi-agent world = genetic algorithms + aggregation + escape. In *7th European Workshop on Modelling Autonomous Agents in a Multi-Agent World (MAA-MAW'96)*. Vrije Universiteit Brussel, Artificial Intelligence Laboratory, 1996.
59. K. Cetnarowicz. M-agent architecture based method of development of multiagent systems. In *8th joint EPS-APS Int. Conf. of Physics Computing*. ACC Cyfronet Kraków, 1996.

60. K. Cetnarowicz. Agent oriented technology based on the m-agent architecture. In *Proceedings of international conference on Intelligent techniques in robotics, control and decision making*, pages 16–31. Polish-Japanese Institute of Information Technology, Warsaw, 1999.

61. K. Cetnarowicz, M. Kisiel-Dorohinicki, and E. Nawarecki. The application of evolution process in multi-agent world (MAW) to the prediction system. In M. Tokoro, editor, *Proc. of the 2nd Int. Conf. on Multi-Agent Systems (ICMAS'96)*. AAAI Press, 1996.

62. S.-H. Chen, Y. Kambayashi, and H. Sato. *Multi-Agent Applications with Evolutionary Computation and Biologically Inspired Technologies*. IGI Global, 2011.

63. C. A. Coello Coello, D. A. Van Veldhuizen, and G. B. Lamont. *Evolutionary Algorithms for Solving Multi-Objective Problems*. Kluwer Academic Publishers, 2002.

64. N. Collier and M. North. *Repast SC++: A Platform for Large-scale Agent-based Modeling*. Wiley, 2011.

65. M. Crepinsek, S.-H. Liu, and M. Mernik. Exploration and exploitation in evolutionary algorithms: A survey. *ACM Computing Surveys*, 45(3), 2013 (in press).

66. I. Crnkovic. *Building Reliable Component-Based Software Systems*. Artech House, 2002.

67. V. Cutello and G. Nicosia. The clonal selection principle for in silico and in vitro computing. In *Recent Developments in Biologically Inspired Computing*, pages 104–146. Idea Group Publishing, 2004.

68. C. Darwin. *The Origin of Species*. Gramercy Books, 1998.

69. D. Dasgupta. *Artificial Immune Systems and Their Applications*. Springer-Verlag, 1998.

70. J. Davila and M. Uzcategui. Galatea: A multi-agent simulation platform. modeling. In *Proc. of Simulation and Neural Networks (MSNN-2000) Merida, Venezuela*, pages 216–233, 2000.

71. T. E. Davis and J. C. Principe. A simulated annealing like convergence theory for the simple genetic algorithm. In *Proc. of the Fourth International Conference on Genetic Algorithms*, pages 174–181, San Diego, CA, 1991.

72. R. Dawkins. *Selfish gene*. Oxford University Press, 1989.

73. L. de Castro and J. Timmis. *Artificial Immune Systems: A New Computational Intelligence Approach*. Springer Verlag, 2002.

74. L. de Castro and J. Timmis. Artificial immune systems: A novel paradigm to pattern recognition. In *Artificial Neural Networks in Pattern Recognition*, pages 67–84. University of Paisley, UK, 2002.

75. L. N. de Castro and F. J. V. Zuben. The clonal selection algorithm with engineering applications. In *Artificial Immune Systems*, pages 36–39, Las Vegas, Nevada, USA, 8 2000.

76. K. De Jong. *An Analysis of the Behavior of a Class of Genetic Adaptive Systems*. PhD thesis, University of Michigan, Ann Arbor, 1975.

77. V. M. de Oliveira, J. F. Fontanari, and P. F. Stadler. Metastable states in short-ranged p-spin glasses. *Journal of Physics A: Mathematical and General*, 32(50):8793, 1999.

78. K. Deb. *Multi-Objective Optimization using Evolutionary Algorithms*. John Wiley & Sons, 2001.

79. Y. Demazeau. *Systèmes multi-agents*. OFTA Paris, 2004.

80. P. D'haeseleer. An immunological approach to change detection: theoretical results. In *Proceedings of the 9th IEEE Computer Security Foundations Workshop*. IEEE, 1996.

81. P. D'haeseleer, S. Forrest, and P. Helman. An immunological approach to change detection: algorithms, analysis and implications. In *Proceedings of the IEEE Symposium on Security and Privacy*. IEEE, 1996.

82. J. Digalakis and K. Margaritis. An experimental study of benchmarking functions for evolutionary algorithms. *International Journal of Computer Mathemathics*, 79(4):403–416, April 2002.

83. G. Dobrowolski and M. Kisiel-Dorohinicki. Management of evolutionary MAS for multi-objective optimization. In T. Burczyński and A. Osyczka, editors, *Evolutionary Methods in Mechanics*, pages 81–90. Kluwer Academic Publishers, 2004.

84. G. Dobrowolski, M. Kisiel-Dorohinicki, and E. Nawarecki. Dual nature of mass multi-agent systems. *Systems Science*, 27(3):77–96, 2002.

85. G. Dobrowolski, M. Kisiel-Dorohinicki, and E. Nawarecki. Agent technology in information systems with explicit knowledge. In Z. Bubnicki and A. Grzech, editors, *Proc. of 15th Int. Conf. on Systems Science*, volume III, pages 105–112. Oficyna Wydawnicza Politechniki Wrocławskiej, Poland, 2004.

86. M. Dorigo. *Optimization, Learning and Natural Algorithms*. PhD thesis, Politecnico di Milano, Italy, 1992.

87. M. Dorigo and T. Stützle. *Ant colony optimization*. MIT Press, 2004.

88. J. Dréo, A. Pétrowski, P. Siarry, E. Taillard, and A. Chatterjee. *Metaheuristics for Hard Optimization: Methods and Case Studies*. Springer, 2005.

89. K. Dresner and P. Stone. A multiagent approach to autonomous intersection management. *Journal of Artificial Intelligence Research*, 31:591–656, 2008.

90. R. Dreżewski. A model of co-evolution in multi-agent system. In V. Mařík, J. Müller, and M. Pěchouček, editors, *Multi-Agent Systems and Applications III*, volume 2691 of *LNCS*, pages 314–323, Berlin, Heidelberg, 2003. Springer-Verlag.

91. R. Dreżewski. Coevolutionary techniques of multi-modal function optimization with application of agent technology (pol. *Koewolucyjne techniki optymalizacji funkcji wielomodalnych z zastosowaniem technologii agentowej*). PhD thesis, Akademia Górniczo-Hutnicza w Krakowie, 2005.

92. R. Dreżewski. Co-evolutionary multi-agent system with speciation and resource sharing mechanisms. *Computing and Informatics*, 25(4):305–331, 2006.

93. R. Dreżewski and K. Cetnarowicz. Sexual selection mechanism for agent-based evolutionary computation. In Y. Shi, G. D. van Albada, J. Dongarra, and P. M. A. Sloot, editors, *Computational Science – ICCS 2007*, volume 4488 of *LNCS*, pages 920–927. Springer-Verlag, 2007.

94. R. Dreżewski and J. Sepielak. Evolutionary system for generating investment strategies. In M. Giacobini, editor, *Applications of Evolutionary Computing*, volume 4974 of *LNCS*, pages 83–92. Springer-Verlag, 2008.

95. R. Dreżewski and L. Siwik. Co-evolutionary multi-agent system with sexual selection mechanism for multi-objective optimization. In *Proc. of IEEE World Congress on Computational Intelligence – WCCI 2006*. IEEE, 2006.

96. R. Dreżewski and L. Siwik. Multi-objective optimization using co-evolutionary multi-agent system with host-parasite mechanism. In V. N. Alexandrov, G. D. van Albada, P. M. A. Sloot, and J. Dongarra, editors, *Computational Science — ICCS 2006*, volume 3993 of *LNCS*, pages 871–878. Springer-Verlag, 2006.

97. R. Dreżewski and L. Siwik. Multi-objective optimization technique based on co-evolutionary interactions in multi-agent system. In M. Giacobini, editor, *Applications of Evolutionary Computing*, volume 4448 of *LNCS*, pages 179–188. Springer-Verlag, 2007.

98. R. Dreżewski and L. Siwik. Agent-based co-operative co-evolutionary algorithm for multiobjective optimization. In L. Rutkowski, R. Tadeusiewicz, L. A. Zadeh, and J. M. Zurada, editors, *Artificial Intelligence and Soft Computing — ICAISC 2008*, volume 5097 of *LNCS*, pages 388–397. Springer-Verlag, 2008.

99. R. Dreżewski and L. Siwik. Agent-based multi-objective evolutionary algorithm with sexual selection. In *Proc. of IEEE World Congress on Computational Intelligence WCCI 2008*. IEEE, 2008.

100. R. Dreżewski and L. Siwik. Co-evolutionary multi-agent system for portfolio optimization. In A. Brabazon and M. O'Neill, editors, *Natural Computation in Computational Finance*, pages 271–299. Springer-Verlag, 2008.

101. R. Dreżewski and L. Siwik. A review of agent-based co-evolutionary algorithms for multiobjective optimization. In Y. Tenne and C.-K. Goh, editors, *Computational Intelligence in Optimization. Aplication and Implementations*. Springer-Verlag, 2010.

102. S. Droste, T. Jansen, and I. Wegener. Upper and lower bounds for randomized search heuristics in black-box optimization. *Theory of Computing Systems*, 39:525–544, 2006.

103. E. H. Durfee and J. Rosenschein. Distributed problem solving and multiagent systems: Comparisons and examples. In M. Klein, editor, *Proceedings of the 13th International Workshop on DAI*, pages 94–104, Lake Quinalt, WA, USA, 1994.

104. R. Eberhart and Y. Shi. *Computational Intelligence: Concepts to Implementations*. Elsevier Science, 2011.

105. N. Eldridge and S. Gould. Punctuated equilibria: An alternative to phyletic gradualism. In T. Schopf, editor, *Models in Paleobiology*. Freeman, Cooper and Co., 1972.

106. J. Farmer, N. Packard, and A. Perelson. The immune system, adaptation, and machine learning. *Physica D: Nonlinear Phenomena*, 22(1–3):187–204, 1986.

107. J. Ferber. Reactive distributed artificial intelligence. In *Foundations of Distributed Artificial Intelligence*, pages 287–317. John Wiley, 1996.

108. J. Ferber. *Multi-Agent Systems. An Introduction to Distributed Artificial Intelligence*. Addison-Wesley, 1999.

109. F. Fernández de Vega and E. Cantú-Paz. *Parallel and Distributed Computational Intelligence*, volume 269 of *Studies in Computational Intelligence*. Springer, 2010.

110. T. Finin, R. Fritzson, D. McKay, and R. McEntire. KQML as an Agent Communication Language. In N. Adam, B. Bhargava, and Y. Yesha, editors, *Proceedings of the 3rd International Conference on Information and Knowledge Management (CIKM'94)*, pages 456–463, Gaithersburg, MD, USA, 1994. ACM Press.

111. FIPA, http://fipa.org/. *The Foundation for Intelligent Physical Agents*.

112. D. B. Fogel. *Evolutionary Computation: Toward a New Philosophy of Machine Intelligence*. IEEE Press, 1995.

113. L. Fogel. Autonomous automata. *Industrial Research*, 4:14–19, 1962.

114. L. Fogel, A. Owens, and M. Walsh. *Artificial Intelligence Through Simulated Evolution*. John Wiley & Sons, New York, 1967.

115. S. Forrest, S. Hofmeyr, and A. Somayaji. Computer immunology. *Communications of the ACM*, 1997.

116. S. Forrest, B. Javornik, R. Smith, and A. Perelson. Using genetic algorithms to explore pattern recognition in the immune system. *Evolutionary Computation*, 1(3):191–211, 1993.

117. S. Forrest and M. Mitchell. Relative building-block fitness and the building-block hypothesis. In D. Whitley, editor, *Foundations of Genetic Algorithms 2*. Morgan Kaufmann, 1993.

118. S. Forrest, A. Perelson, L. Allen, and R. Cherukuri. Self-nonself discrimination in a computer. In *Proceedings of the 1992 IEEE Symposium on Security and Privacy*. IEEE, 1994.

119. M. Fowler. *Domain-Specific Languages*. Addison-Wesley, 2010.

120. R. Friedberg, B. Dunham, and T. North. A learning machine: Part II. *IBM Journal of Research and Development*, 3(3), 1959.

121. R. M. Friedberg. A learning machine: Part I. *IBM Journal of Research and Development*, 2(1), 1958.

122. R. Gagliardi, J. Robbins, and H. Taylor. Acquisition sequences in ppm communications (corresp.). *Information Theory, IEEE Transactions on*, 33(5):738–744, 1987.

123. J. E. Gallardo, C. Cotta, and A. J. Fernández. Finding low autocorrelation binary sequences with memetic algorithms. *Applied Soft Computing*, 9(4):1252–1262, 2009.

124. E. Gamma, R. Helm, R. Johnson, and J. Vlissides. *Design Patterns: Elements of Reusable Object-Oriented Software*. Addison-Wesley, 1995.

125. R. Garello, N. Boujnah, and Y. Jia. Design of binary sequences and matrices for space applications. In *Satellite and Space Communications, 2009. IWSSC 2009. International Workshop on*, pages 88–91, 2009.

126. M. Genesereth and N. Nilsson. *Logical Foundations of Artificial Intelligence*. Morgan Kaufmann Publishers, San Mateo, CA, 1987.

127. M. R. Genesereth and S. P. Ketchpel. Software agents. *Communications of the ACM*, 37(7):48–53, 1994.

128. J. George, M. Gleizes, P.Glize, and C.Regis. Real-time simulation for flood forecast: an adaptive multi-agent system staff. In *Proceedings of the AISB'03 Symposium on Adaptive Agents and Multi-Agent Systems*. University of Wales, 2003.

129. F. Glover. Future paths for integer programming and links to artificial intelligence. *Comput. Oper. Res.*, 13(5):533–549, May 1986.

130. F. Glover and K. G.A. *Handbook of Metaheuristics*. Springer, 2003.

131. F. Glover and C. McMillan. The general employee scheduling problem: an integration of ms and ai. *Comput. Oper. Res.*, 13(5):563–573, May 1986.
132. K. Gödel. Über formal unentscheidbare Sätze der Principia Mathematica und verwandter Systeme I. *Monatsheft für Math. und Physik*, 38:173–198, 1931.
133. D. E. Goldberg. *Genetic Algorithms and their applications*. Pearson, 1996.
134. D. E. Goldberg and P. Segrest. Finite Markov chain analysis of genetic algorithms. In *Proceedings of the Second International Conference on Genetic Algorithms on Genetic algorithms and their application*, pages 1–8, Hillsdale, NJ, USA, 1987. L. Erlbaum Associates Inc.
135. E. Grabska, K. Grzesiak-Kopeć, and G. Ślusarczyk. Designing floor-layouts with the assistance of curious agents. In V. N. Alexandrov, G. D. van Albada, P. M. A. Sloot, and J. Dongarra, editors, *International Conference on Computational Science (3)*, volume 3993 of *Lecture Notes in Computer Science*, pages 883–886. Springer, 2006.
136. E. Grabska, K. Grzesiak-Kopeć, and G. Ślusarczyk. Visual creative design with the assistance of curious agents. In D. Barker-Plummer, R. Cox, and N. Swoboda, editors, *Diagrams*, volume 4045 of *Lecture Notes in Computer Science*, pages 218–220. Springer, 2006.
137. J. Grefenstette. Optimization of control parameters for genetic algorithms. *IEEE Trans. on Systems, Man and Cybernetics*, 16(1):122–128, 1986.
138. M. Grochowski, R. Schaefer, and P. Uhruski. Diffusion based scheduling in the agent-oriented computing systems. *Lecture Notes in Computer Science*, 3019:97–104, 2004.
139. O. Gutknecht and J. Ferber. The madkit agent platform architecture. In T. Wagner and O. F. Rana, editors, *Infrastructure for Agents, Multi-Agent Systems, and Scalable Multi-Agent Systems*, volume 1887 of *LNCS*, pages 48–55. Springer, 2001.
140. P. Hajela and J. Lee. Constrained genetic search via schema adaptation. an immune network solution. *Structural optimization*, 12:11–15, 1996.
141. S. Halim, R. H. Yap, and F. Halim. Engineering stochastic local search for the low autocorrelation binary sequence problem. In *Principles and Practice of Constraint Programming*, pages 640–645. Springer, 2008.
142. W. Hastings. Monte Carlo sampling methods using markov chains and their applications. *Biometrika*, 57(1), 1970.
143. J. He and X. Yao. From an individual to a population: An analysis of the first hitting time of population-based evolutionary algorithms. *IEEE Trans. on Evolutionary Computation*, 6(5):495–511, 2002.
144. G. Hinton and S. Nolan. How learning can guide evolution. *Complex Systems*, 1:495–502, 1987.
145. C. Hoare. Communicating sequential processes. *Commun. ACM*, 21(8):666–677, Aug. 1978.
146. G. Hoffmann. A neural network model based on the analogy with the immune system. *Journal of Theorectical Biology*, 122(1):33–67, 1986.
147. J. Holland. *Adaptation in natural and artificial systems*. MIT Press, 1975.
148. J. H. Holland. Outline for a logical theory of adaptive systems. *Journal of Assoc. for Computing Machinery*, 9(3), 1962.
149. H. Hoos and T. Stützle. *Stochastic Local Search: Foundations and Applications*. Morgan Kaufmann, 2004.
150. R. Horst and P. Pardalos. *Handbook of Global Optimization*. Kluwer, 1995.
151. M. Iosifescu. *Finite Markov Processes and Their Applications*. John Wiley and Sons, 1980.
152. Y. Ishida. Fully distributed diagnosis by pdp learning algorithm: towards immune network pdp models. In *IJCNN International Joint Conference on Neural Networks*. IEEE, 1990.
153. N. Jennings, P. Faratin, M. Johnson, T. Norman, P. OBrien, and M. Wiegand. Agent-based business process management. *International Journal of Cooperative Information Systems*, 5(2–3):105–130, 1996.
154. N. R. Jennings, K. Sycara, and M. Wooldridge. A roadmap of agent research and development. *Journal of Autonomous Agents and Multi-Agent Systems*, 1(1):7–38, 1998.
155. N. R. Jennings and M. J. Wooldridge. Software agents. *IEE Review*, pages 17–20, 1996.
156. N. R. Jennings and M. J. Wooldridge. Applications of intelligent agents. In N. R. Jennings and M. J. Wooldridge, editors, *Agent Technology: Foundations, Applications, and Markets*, pages 3–28. Springer Verlag: Heidelberg, Germany, 1998.

157. R. E. Johnson and B. Foote. Designing Reusable Classes. *Object-Oriented Programming*, 1, 1988.
158. A. J. F. Jos E. Gallardo, Carlos Cotta. Finding low autocorrelation binary sequences with memetic algorithms. *Applied Soft Computing*, 9, 2009.
159. A. Juels and M. Wattenberg. Stochastic hillclimbing as a baseline method for evaluating genetic algorithms. Technical report, University of California at Berkeley, 1994.
160. L. Jun-qing, P. Quan-ke, and L. Yun-Chia. An effective hybrid tabu search algorithm for multi-objective flexible job-shop scheduling problems. *Computers & Industrial Engineering*, 59(4):647–662, 2010.
161. A. Kay. Computer software. *Scientific American*, 251(3):53–59, 1990.
162. J. Kelsey and J. Timmis. Immune inspired somatic contiguous hypermutation for function optimization. In E. Cantú-Paz, editor, *GECCO Genetic and Evolutionary Computation Conference, LNCS Vol. 2723*, pages 207–218. Springer, 2003.
163. J. Kennedy and R. Eberhart. Particle swarm optimization. In *Proc. of IEEE Int. Conf. on Neural Networks*, volume 4, pages 1942–1948, 1995.
164. J. Kennedy, J. Kennedy, R. Eberhart, and Y. Shi. *Swarm intelligence*. Morgan Kaufmann Publishers, 2001.
165. J. Kephart. A biologically inspired immune system for computers. In *Artificial Life IV*. MIT Press, 1994.
166. J. Kephart, G. Sorkin, W. Arnold, D. M. Chess, G. Tesauro, and S. White. Biologically inspired defences against computer viruses. In *Proceedings of the 14th International Joint Conference on Artificial Intelligence*. IJCAI, 1995.
167. D. Kinny and M. Georgeff. Modelling and design of multi-agent systems. In Müller et al. [220].
168. S. Kirkpatrick, C. D. Gelatt, and M. P. Vecchi. Optimization by simulated annealing. *Science*, 220:671–680, 1983.
169. M. Kisiel-Dorohinicki. Agent-oriented model of simulated evolution. In W. I. Grosky and F. Plasil, editors, *SofSem 2002: Theory and Practice of Informatics*, volume 2540 of *LNCS*. Springer-Verlag, 2002.
170. M. Kisiel-Dorohinicki. Mechanism of 'crowd' in evolutionary MAS for multiobjective optimisation. In J. Arabas, editor, *Evolutionary Computation and Global Optimization*, volume 139 of *Prace Naukowe, Elektronika*. Oficyna Wydawnicza Politechniki Warszawskiej, 2002.
171. M. Kisiel-Dorohinicki. Platform for data exchange in a distributed heterogeneous network environment (pol. platforma wymiany wiedzy w rozproszonym heterogenicznym środowisku sieciowym). Technical Report 4/2003, Katedra Informatyki, Akademia Górniczo-Hutnicza w Krakowie, 2003.
172. M. Kisiel-Dorohinicki. Agent-based models and platforms for parallel evolutionary algorithms. In M. Bubak, G. D. van Albada, P. M. A. Sloot, and J. Dongarra, editors, *Computational Science – ICCS 2004. Proc. of 4th Int. Conf. Part III*, volume 3038 of *LNAI*, pages 225–236. Springer-Verlag, 2004.
173. M. Kisiel-Dorohinicki. Flock-based architecture for distributed evolutionary algorithms. In L. Rutkowski, J. Siekmann, R. Tedeusiewicz, and L. Zadeh, editors, *Artificial Intelligence and Soft Computing – ICAISC 2004. Proc. of 7th Int. Conf.*, volume 3070 of *LNAI*, pages 841–846. Springer-Verlag, 2004.
174. M. Kisiel-Dorohinicki. Simulation of mechanisms of species competition in an environment with limited resources (pol. symulacja mechanizmów konkurencji gatunkowej w środowisku o ograniczonych zasobach). Technical Report 3/2004, Katedra Informatyki, Akademia Górniczo-Hutnicza w Krakowie, 2004.
175. M. Kisiel-Dorohinicki. *Agent-based architectures of computing systems (in Polish: Agentowe architektury populacyjnych systemów obliczeniowych)*. Dissertations and monographs, vol. 269, AGH University of Science and Technology Press, 2013.
176. M. Kisiel-Dorohinicki, G. Dobrowolski, and E. Nawarecki. Evolutionary multi-agent system in multiobjective optimisation. In M. Hamza, editor, *Proc. of IASTED Int. Symp.: Applied Informatics*. IASTED/ACTA Press, 2001.

177. M. Kisiel-Dorohinicki, G. Dobrowolski, and E. Nawarecki. Agent populations as computational intelligence. In L. Rutkowski and J. Kacprzyk, editors, *Neural Networks and Soft Computing*, pages 608–614. Physica Verlag, 2002.

178. M. Kisiel-Dorohinicki and K. Socha. Crowding factor in evolutionary multi-agent system for multiobjective optimization. In H. R. Arabnia, editor, *Proc. of Int. Conf. on Artificial Intelligence (IC-AI 2001)*. CSREA Press, 2001.

179. P. Koch, T. Simpson, J. Allen, and F. Mistree. Statistical approximations for multidisciplinary design optimization: the problem of size. *Journal of Aircraft*, 36(1):275–286, 1999.

180. J. Kołodziej. A simple markov model and asymptotic properties of a hierarchical genetic strategy. *Schedae Informaticae*, 11:41–55, 2002.

181. J. R. Koza. *Genetic Programming: On the Programming of Computers by Means of Natural Selection*. Cambridge, MA: MIT Press, 1992.

182. N. Krasnogor and S. Gustafson. A study on the use of "self-generation" in memetic algorithms. *Natural Computing*, 3:53–76, 2004.

183. N. Krasnogor and J. Smith. A tutorial for competent memetic algorithms: Model, taxonomy, and design issues. *IEEE Transactions on Evolutionary Computation*, 9(5):474–488, 2005.

184. H. Kreger, W. Harold, and L. Williamson. *Java and JMX: building manageable systems*. Addison-Wesley, 2003.

185. R. Krutisch, P. Meier, and M. Wirsing. The agent–component approach, combining agents and components. In M. Schillo, M. Klusch, J. P. Müller, and H. Tianfield, editors, *Proc. of 1st German Conf. on Multiagent System Technologies, MATES 2003*, volume 2831 of *LNCS*. Springer, 2003.

186. D. Krzywicki. Niching in evolutionary multi-agent systems. *Computer Science*, 14(1), 2013.

187. K. Ku and M. Mak. Exploring the effects of lamarckian and baldwinian learning in evolving recurrent neural networks. In *Proc. of 1997 IEEE Int. Conf. on Evolutionary Computation*. IEEE, 1997.

188. J.-S. Lee and C. H. Park. Hybrid simulated annealing and its application to optimization of hidden markov models for visual speech recognition. *Systems, Man, and Cybernetics, Part B: Cybernetics, IEEE Transactions on*, 40(4):1188–1196, aug. 2010.

189. W. D. Li, S. K. Ong, and A. Y. C. Nee. Hybrid genetic algorithm and simulated annealing approach for the optimization of process plans for prismatic parts. *International Journal of Production Research*, 40(8):1899–1922, 2002.

190. B. H. Liskov and J. M. Wing. A behavioral notion of subtyping. *ACM Trans. on Programming Languages and Systems*, 16(6):1811–1841, 1994.

191. B. Lobel, A. Ozdaglar, and D. Feijer. Distributed multi-agent optimization with state-dependent communication. *Mathematical Programming*, 129(2):255–284, 2011.

192. M. Locatelli. Simulated annealing algorithms for continuous global optimization: Convergence conditions. *Journal of Optimization Theory and Applications*, 104:121–133, 2000.

193. M. Luban and L. P. Staunton. An efficient method for generating a uniform distribution of points within a hypersphere. *Computers in Physics*, 2:55–60, Nov. 1988.

194. S. Luke, C. Cioffi-Revilla, L. Panait, K. Sullivan, and G. Balan. MASON: A multi-agent simulation environment. *Simulation: Trans. of Society for Modeling and Simulation Int.*, 82(7):517–527, 2005.

195. P. Maes. Agents that reduce work and information overload. *Communications of the ACM*, 37(7):30–40, 1987.

196. P. Mahalanobis. On the generalised distance in statistics. *Proceedings of the National Institute of Sciences of India*, 2(1):49–55, 1936.

197. S. Mahfoud. Finite Markov Chain Models of an Alternative Selection Strategy for the Genetic Algorithm. *Complex Systems*, 7:155–170, 1991.

198. S. W. Mahfoud. A comparison of parallel and sequential niching methods. In *In Proceedings of the Sixth International Conference on Genetic Algorithms*, pages 136–143. Morgan Kaufmann, 1995.

199. J. March. Exploration and exploitation in organizational learning. *Organization Science*, 2:71–87, 1991.

200. G. Martin. Optimal convolutional self-orthogonal codes with an application to digital radio. In *ICC'85; International Conference on Communications*, volume 1, pages 1249–1253, 1985.

201. R. C. Martin. The dependency inversion principle. *C++ Report*, 8(6):61–66, 1996.

202. R. C. Martin. *Agile Software Development: Principles, Patterns, and Practices*. Prentice Hall, 2003.

203. W. Martin, J. Lienig, and J. Cohoon. Island (migration) models: evolutionary algorithms based on punctuated equilibria. In T. Bäck, D. B. Fogel, and Z. Michalewicz, editors, *Handbook of Evolutionary Computation*. IOP Publishing and Oxford University Press, 1997.

204. M. H. Mashinchi, M. A. Orgun, and W. Pedrycz. Hybrid optimization with improved tabu search. *Applied Soft Computing*, 11(2):1993–2006, 2011. The Impact of Soft Computing for the Progress of Artificial Intelligence.

205. J. McAffer, J. Lemieux, and C. Aniszczyk. *Eclipse Rich Client Platform*. Pearson Education, 2010.

206. J. McAffer, P. VanderLei, and S. Archer. *OSGi and Equinox: Creating Highly Modular Java Systems*. Addison-Wesley Professional, 2010.

207. S. McArthur, V. Catterson, and N. Hatziargyriou. Multi-agent systems for power engineering applicationspart i: Concepts, approaches, and technical challenges. *IEEE TRANSACTIONS ON POWER SYSTEMS*, 22(4), November 2007.

208. M. McIlroy. Mass-produced software components. *Proc. of NATO Conf. on Software Engineering*, 1968.

209. N. McPhee and N. Hopper. Analysis of genetic diversity through population history. In *Proceedings of the Genetic and Evolutionary Computation Conference (GECCO), 13-17 July, Orlando FL, USA*, pages 1112–1120, 1999.

210. Z. Michalewicz. *Genetic Algorithms + Data Structures = Evolution Programs*. Springer-Verlag, 1996.

211. Z. Michalewicz. Ubiquity symposium: Evolutionary computation and the processes of life: the emperor is naked: evolutionary algorithms for real-world applications. *Ubiquity*, 2012(November):3:1–3:13, Nov. 2012.

212. Z. Michalewicz and D. Fogel. *How to Solve It: Modern Heuristics*. Springer, 2004.

213. R. Michalski. Learnable evolution model: Evolutionary processes guided by machine learning. *Machine Learning*, 38:9–40, 2000.

214. N. Minar, R. Burkhart, C. Langton, and M. Askenazi. The swarm simulation system: A toolkit for building multi-agent simulations. Technical report, Santa Fe Institute, 1996.

215. M. Mitchell, J. H. Holland, and S. Forrest. When will a genetic algorithm outperform hill climbing? In *Advances in Neural Information Processing Systems 6*, pages 51–58. Morgan Kaufmann, 1993.

216. A. Moffet. Minimum-redundancy linear arrays. *Antennas and Propagation, IEEE Transactions on*, 16(2):172–175, 1968.

217. R. W. Morrison and K. A. D. Jong. Measurement of population diversity. In *In 5th International Conference EA 2001*, pages 31–41. Springer, 2002.

218. P. Moscato. Memetic algorithms: a short introduction. In *New ideas in optimization*, pages 219–234, Maidenhead, UK, England, 1999. McGraw-Hill Ltd., UK.

219. P. Moscato and C. Cotta. A modern introduction to memetic algorithms. In M. Gendrau and J.-Y. Potvin, editors, *Handbook of Metaheuristics*, volume 146 of *Int. Series in Operations Research and Management Science*, pages 141–183. Springer, 2 edition, 2010.

220. J. P. Müller, M. Wooldridge, and N. R. Jennings, editors. *Intelligent Agents III: Proc. of ECAI'96 Workshop on Agent Theories, Architectures, and Languages*, volume 1193 of *LNAI*. Springer-Verlag, 1997.

221. E. Nawarecki, M. Kisiel-Dorohinicki, and G. Dobrowolski. Organisations in the particular class of multi-agent systems. In B. Dunin-Keplicz and E. Nawarecki, editors, *From Theory to Practice in Multi Agent Systems*, volume 2296 of *LNAI*. Springer-Verlag, 2002.

222. F. Neri. Diversity management in memetic algorithms. In F. Neri, C. Cotta, and P. Moscato, editors, *Handbook of Memetic Algorithms*, volume 379 of *Studies in Computational Intelligence*, pages 153–165. Springer Berlin Heidelberg, 2012.

223. C. Nikolai and G. Madey. Tools of the trade: A survey of various agent based modeling platforms. *Journal of Artificial Societies and Social Simulation*, 12(2), 2008.
224. A. E. Nix and M. D. Vose. Modeling genetic algorithms with markov chains. *Annals of Mathematics and Artificial Intelligence*, 5(1):79–88, 1992.
225. M. North, T. Howe, N. Collier, and J. Vos. A declarative model assembly infrastructure for verification and validation. In S. Takahashi, D. Sallach, and J. Rouchier, editors, *Advancing Social Simulation: 1st World Congress*. Springer, 2007.
226. OMG Unified Modeling Language, Superstructure Specification, Version 2.4.1. Technical Report formal/2011-08-06, Object Management Group, 2011.
227. Y.-S. Ong, M.-H. Lim, N. Zhu, and K.-W. Wong. Classification of adaptive memetic algorithms: a comparative study. *Systems, Man, and Cybernetics, Part B: Cybernetics, IEEE Transactions on*, 36(1):141–152, 2006.
228. A. Osyczka. *Evolutionary Algorithms for Single and Multicriteria Design Optimization*. Physica Verlag, 2002.
229. C. Papadimitriou and K. Steiglitz. *Combinatorial Optimization: Algorithms and Complexity*. Dover Publications, Inc., 1998.
230. J. Paredis. Coevolutionary computation. *Artificial Life*, 2(4):355–375, 1995.
231. D. L. Parnas. On the criteria to be used in decomposing systems into modules. *Communications ACM*, 15(12), 1972.
232. C. Pettey. Diffusion (cellular) models. In T. B"ack, D. B. Fogel, and Z. Michalewicz, editors, *Handbook of Evolutionary Computation*. IOP Publishing and Oxford University Press, 1997.
233. K. Pietak and M. Kisiel-Dorohinicki. Agent-based framework facilitating component-based implementation of distributed computational intelligence systems. *Trans. on Computational Collective Intelligence*, 10, 2013.
234. K. Pietak, A. Woś, A. Byrski, and M. Kisiel-Dorohinicki. Functional integrity of multi-agent computational system supported by component-based implementation. In *Proc. of 4th Int. Conf. on Industrial Applications of Holonic and Multi-agent Systems*, volume 5696 of *LNAI*. Springer-Verlag, 2009.
235. S. Pisarski, A. Rugała, A. Byrski, and M. Kisiel-Dorohinicki. Evolutionary multi-agent system in hard benchmark continuous optimisation. In *Proc. of EVOSTAR Conference, Vienna*. IEEE (accepted for printing), 2013.
236. M. Potter and K. De Jong. Cooperative coevolution: An architecture for evolving coadapted subcomponents. *Evolutionary Computation*, 8(1):1–29, 2000.
237. S. Railsback and L. Lytinen. Agent-based simulation platforms: review and development recommendations. *Simulations*, 82:609–623, 2006.
238. A. S. Rao and M. P. Georgeff. Bdi agents: From theory to practice. In *First International Conference on Multi-Agent Systems (ICMAS-95*, pages 312–319. AAAI, 1995.
239. I. Rechenberg. Cybernetic solution path of an experimental problem (kybernetische Lösungsansteuerung einer experimentellen Forschungsaufgabe). Royal Aircraft Establishment, Library translation No. 1122, 1965. presented at the Annual Conf. of WGLR, Berlin, 1964.
240. I. Rechenberg. *Evolutionstrategie: Optimierung technischer systeme nach prinzipien der biologischen evolution*. Formann-Holzboog, 1973.
241. A. Rinnoy Kan and G. Timmer. Stochastic global optimization methods. *Mathematical Programming*, 39:27–56, 1987.
242. J. Robinson and A. Bernstein. A class of binary recurrent codes with limited error propagation. *Information Theory, IEEE Transactions on*, 13(1):106–113, 1967.
243. S. Robinson. *Simulation: The Practice of Model Development and Use*. Wiley, 2004.
244. J. Rowe, K. Vinsen, and N. Marvin. Parallel GAs for multiobjective functions. In J. T. Alander, editor, *Proc. of 2nd Nordic Workshop on Genetic Algorithms and Their Applications (2NWGA)*. University of Vaasa, Finnland, 1996.
245. G. Rudolph. Convergence analysis of canonical genetic algorithms. *IEEE Transactions on Neural Networks*, 5(1):96–101, 1994.

246. G. Rudolph. Evolution strategies. In T. Bäck, D. Fogel, and Z. Michalewicz, editors, *Handbook of Evolutionary Computations*. Oxford University Press, 1997.

247. G. Rudolph. Models of stochastic convergence. In T. Bäck, D. Fogel, and Z. Michalewicz, editors, *Handbook of Evolutionary Computations*. Oxford University Press, 1997.

248. G. Rudolph. Stochastic processes. In T. Bäck, D. Fogel, and Z. Michalewicz, editors, *Handbook of Evolutionary Computations*. Oxford University Press, 1997.

249. G. Rudolph. Finite Markov chain results in evolutionary computation: A tour d'horizon. *Fundamenta Informaticae*, 35(1–4):67–89, 1998.

250. S. Russel and P. Norvig. *Artificial Intelligence: A Modern Approach*. Prentice Hall, 2003.

251. D. Rutkowska. *Neuro-Fuzzy Architectures and Hybrid Learning*, volume 85 of *Studies in Fuzziness and Soft Computing*. Physica-Verlag, 2002.

252. R. Sarker and T. Ray. *Agent-Based Evolutionary Search*, volume 5 of *Adaptation, Learning and Optimization*. Springer, 1 edition, 2010.

253. R. Schaefer. *Foundations of global genetic optimization*. Springer Verlag, 2007.

254. R. Schaefer, A. Byrski, J. Kołodziej, and M. Smołka. An agent-based model of hierarchic genetic search. *Computers and Mathematics with Applications*, 64(12):3763–3776, 2012.

255. R. Schaefer, A. Byrski, and M. Smołka. Stochastic model of evolutionary and immunological multi-agent systems: Parallel execution of local actions. *Fundamenta Informaticae*, 95(2-3):325–348, 2009.

256. R. Schaefer, A. Byrski, and M. Smołka. The island model as a markov dynamic system. *International Journal of Applied Mathematics and Computer Science*, 12(4), 2012.

257. R. Schaefer, W. Toporkiewicz, and M. Grochowski. Rough partitioning of lumped structures. In *Formal Methods and Intelligent Techniques in Control, Decision Making, Multimedia and Robotics*, pages 151–166. Polish-Japanese Institute of Information Technology Press, Oct. 2000.

258. H. Schwefel. *Evolution and optimum seeking*. Chichester, Wiley, 1995.

259. H.-P. Schwefel. Kybernetische evolution als strategie der experimentellen forschung inder strömungstechnik. Technical report, Technische Universität, Berlin, 1965.

260. H.-P. Schwefel. *Evolutionsstrategie und numerische Optimierung*. PhD thesis, Technische Universität Berlin, Berlin, Germany, 1975.

261. J. B. Shearer. Some new optimum golomb rulers. *Information Theory, IEEE Transactions on*, 36(1):183–184, 1990.

262. L. Siwik. *Agent-based methods for improving efficiency of the evolutionary multi-criteria optimization techniques (pol. Agentowe metody poprawy efektywności ewolucyjnych technik optymalizacji wielokryterialnej)*. PhD thesis, Akademia Górniczo-Hutnicza w Krakowie, 2009.

263. L. Siwik and R. Dreżewski. Agent-based multi-objective evolutionary algorithms with cultural and immunological mechanisms. In W. P. dos Santos, editor, *Evolutionary computation*, pages 541–556. In-Teh, 2009.

264. L. Siwik and M. Kisiel-Dorohinicki. Elitism in agent-based evolutionary multiobjective optimization. In *Inteligencia Artificial, Special issue: New trends on Multiagent systems and soft computing* [1], pages 41–48.

265. L. Siwik and M. Kisiel-Dorohinicki. Semi-elitist evolutionary multi-agent system for multiobjective optimization. In V. N. Alexandrov, G. D. van Albada, P. Sloot, and J. Dongara, editors, *Computational Science – ICCS 2006: Proc. of 6th Int. Conf.*, volume 3993 of *LNCS*. Springer-Verlag, 2006.

266. L. Siwik and M. Kisiel-Dorohinicki. Improving the quality of the pareto frontier approximation obtained by semi-elitist evolutionary multi-agent system using distributed and decentralized frontier crowding mechanism. In *Adaptive and natural computing algorithms – ICAN-NGA 2007, Proc. of 8th Int. Conf.*, volume 4431 of *LNCS*. Springer-Verlag, 2007.

267. A. Skulimowski. *Decision Support Systems Based on Reference Sets*. AGH Publishing House, Krakow, Poland, 1996.

268. K. Socha and M. Kisiel-Dorohinicki. Agent-based evolutionary multiobjective optimisation. In *Proc. of 2002 Congress on Evolutionary Computation*. IEEE, 2002.

269. F. Solis and R. Wets. Minimization by random search techniques. *Mathematical Methods of Operations Research*, 6:19–30, 1981.

270. K. Sörensen. Metaheuristics—the metaphor exposed. *International Transactions in Operational Research*, 22(1):3–18, 2015.

271. P. F. Stadler. Landscapes and their correlation functions. *Journal of Mathematical chemistry*, 20(1):1–45, 1996.

272. I. Stadnyk. Schema recombination in a pattern recognition process. In *Proc. of the Second Int. Conf. on Genetic Algorithms*, pages 27–35. Cambridge, MA: Lawrence Erlbaum Associates, 1987.

273. R. Steur. *Multiple Criteria Optimization: Theory, Computation and Application*. John Willey & Sons, 1986.

274. J. Suzuki. A Markov Chain Analysis on a Genetic Algorithm. In S. Forrest, editor, *Proceedings of the 5th International Conference on Genetic Algorithms, Urbana-Champaign, IL, USA, June 1993*, pages 146–154. Morgan Kaufmann, 1993.

275. R. E. Sweet. The mesa programming environment. In *Proc. of ACM SIGPLAN 85 symposium on Language issues in programming environments*, SLIPE '85, pages 216–229. ACM, 1985.

276. C. Szyperski. *Component Software: Beyond Object-Oriented Programming*. Addison-Wesley Longman, 2002.

277. E.-G. Talbi. A taxonomy of hybrid metaheuristics. *Journal of Heuristics*, 8(5):541–564, 2002.

278. E.-G. Talbi. *Metaheuristics: From Design to Implementation*. Wiley, 2009.

279. N. Talbi and K. Belarbi. Optimization of fuzzy controller using hybrid tabu search and particle swarm optimization. In *Hybrid Intelligent Systems (HIS), 2011 11th International Conference on*, pages 561 –565, dec. 2011.

280. M. Tomassini. *Spatially Structured Evolutionary Algorithms: Artificial Evolution in Space and Time*. Natural Computing Series. Springer, 2010.

281. R. Turyn. Sequences with small correlation. *Error correcting codes*, pages 195–228, 1968.

282. R. Turyn and J. Storer. On binary sequences. *Proceedings of the American Mathematical Society*, 12(3):394–399, 1961.

283. A. M. Uhrmacher and D. Weyns. *Multi-agent systems: simulation and applications*. CRC Press, 2009.

284. P. Uhruski, M. Grochowski, and R. Schaefer. Multi-agent computing system in a heterogeneous network. In *Proceedings of the International Conference on Parallel Computing in Electrical Engineering (PARELEC 2002)*, pages 233–238, Warsaw, Poland, 22–25 Sept. 2002. IEEE Computer Society Press.

285. P. Uhruski, M. Grochowski, and R. Schaefer. Octopus – computation agents environment. *Inteligencia Artificial, Revista Iberoamericana de IA*, 9(28):55–62, 2005.

286. P. Uhruski, M. Grochowski, and R. Schaefer. A two-layer agent-based system for large-scale distributed computation. *Computational Intelligence*, 24(3):191–212, Aug. 2008.

287. A. Ukil. Low autocorrelation binary sequences: Number theory-based analysis for minimum energy level, barker codes. *Digital Signal Processing*, 20(2):483–495, 2010.

288. T. Victoire and A. Jeyakumar. Unit commitment by a tabu-search-based hybrid-optimisation technique. *Generation, Transmission and Distribution, IEE Proceedings-*, 152(4):563–574, july 2005.

289. T. A. A. Victoire and A. E. Jeyakumar. A tabu search based hybrid optimization approach for a fuzzy modelled unit commitment problem. *Electric Power Systems Research*, 76(6–7):413–425, 2006.

290. M. Vose. *The Simple Genetic Algorithm: Foundations and Theory*. MIT Press, Cambridge, MA, USA, 1998.

291. M. D. Vose and G. E. Liepins. Punctuated eqlibria in genetic search. *Complex Systems*, 5(1):31–44, 1991.

292. H. Wang, D. Wang, and S. Yang. A memetic algorithm with adaptive hill climbing strategy for dynamic optimization problems. *Soft Computing*, 13(8–9):763–780, 2008.

293. D. Whitley. An executable model of a simple genetic algorithm. In *Foundations of Genetic Algorithms 2*, pages 45–62. Morgan Kaufmann, 1992.

294. D. Whitley, V. Scott Gordon, and K. Mathias. Lamarckian evolution, the baldwin effect and function optimization. In Davidor, Y. and Schwefel, H.-P. and Männer, R., editor, *Proc. of Parallel Problem Solving from Nature III*. Springer, 1994.

295. S. Wierzchoń. *Artificial immune systems. Theory and applications. (pol. Sztuczne systemy immunologiczne. Teoria i zastosowania)*. EXIT, 2001.

296. D. Wolpert and W. Macready. No free lunch theorems for search. Technical Report SFI-TR-02-010, Santa Fe Institute, 1995.

297. D. Wolpert and W. Macready. No free lunch theorems for optimization. *IEEE Transactions on Evolutionary Computation*, 67(1), 1997.

298. M. Wooldridge. Agent-based software engineering. *IEEE Trans. on Software Engineering*, 144(1), 1997.

299. M. Wooldridge. *An Introduction to Multiagent Systems*. John Wiley & Sons, 2009.

300. M. Wooldridge and N. Jennings. Intelligent agents. In *LNAI 890*. Springer Verlag, 1995.

301. M. Wooldridge and N. R. Jennings. Pitfalls of agent-oriented development. In K. P. Sycara and M. Wooldridge, editors, *Proc. of 2nd Int. Conf. on Autonomous Agents (Agents'98)*. ACM Press, 1998.

302. X. Xu and H.-g. He. A theoretical model and convergence analysis of memetic evolutionary algorithms. In L. Wang, K. Chen, and Y. S. Ong, editors, *Advances in Natural Computation*, volume 3611 of *Lecture Notes in Computer Science*, pages 1035–1043. Springer Berlin / Heidelberg, 2005.

303. X. Yao, F. Wang, K. Padmanabhan, and S. Salcedo-Sanz. Hybrid evolutionary approaches to terminal assignment in communications networks. In W. Hart, J. Smith, and N. Krasnogor, editors, *Recent Advances in Memetic Algorithms*, volume 166 of *Studies in Fuzziness and Soft Computing*, pages 129–159. Springer Berlin Heidelberg, 2005.

304. W. Zhong, J. Liu, M. Xue, and L. Jiao. A multiagent genetic algorithm for global numerical optimization. *IEEE Trans. on Systems, Man, and Cybernetics, Part B: Cybernetics*, 34(2):1128–1141, 2004.

305. Y. Zhong and X. Pan. A hybrid optimization solution to vrptw based on simulated annealing. In *Automation and Logistics, 2007 IEEE International Conference on*, pages 3113–3117, aug. 2007.

Printed in the United States
By Bookmasters